T0351128

The Cause, Effect, and Control of Accidental Loss

The Cause, Effect, and Control of Accidental Loss takes the reader through 15 phases of a typical workplace accident and shows how accidents can be prevented by the introduction of safety management controls in the form of a structured health and safety management system (SMS). It proposes that once the event has been triggered, there is no certainty as to the outcome, so workplaces should rely on proactive safety actions rather than reactive ones. Now fully updated, this new edition expands on the important concepts from the first edition, including hazard identification, risk assessment, flawed safety management systems, the potential for loss, and management control.

This title:

- Challenges the paradigm that the measure of consequence (losses) is a good indicator of safety effort.
- Introduces three luck factors that determine the course of the accident sequence.
- Explains what causes accidents, their consequences, and how to prevent them.
- Showcases accident immediate causes including high-risk (unsafe) acts and high-risk (unsafe) conditions.

The text is an essential read for professionals, graduate students, and academics in the field of occupational health, safety, and industrial hygiene.

Workplace Safety, Risk Management, and Industrial Hygiene

Series Editor
Ron C. McKinnon, McKinnon & Associates,
Western Cape, South Africa

This new series will provide the reader with a comprehensive selection of publications covering all topics appertaining to the health and safety of people at work, at home, and during recreation. It will deal with the technical aspects of the profession, as well as the psychological ramifications of safe human behavior. This new series will include books covering the areas of accident prevention, loss control, physical risk management, safety management systems, occupational safety, industrial hygiene, occupational medicine, public safety, home safety, recreation safety, safety management, injury management, near miss incident management, school safety guidance, and other related areas within the Occupational Health and Safety discipline.

Safety Management: Near Miss Identification,
Recognition, and Investigation
Ron C. McKinnon

Changing the Workplace Safety Culture
Ron C. McKinnon

Risk-based, Management-led, Audit-driven, Safety Management Systems
Ron C. McKinnon

The Design, Implementation, and Audit of Occupational
Health and Safety Management Systems
Ron C. McKinnon

Cause, Effect, and Control of Accidental Loss
with Accident Investigation Kit
Ron C. McKinnon

A Practical Guide to Effective Workplace Accident Investigation
Ron C. McKinnon

The Cause, Effect, and Control of Accidental Loss, Second Edition
Ron C. McKinnon

For more information on this series, please visit: https://www.crcpress.com/ Workplace-Safety-Risk-Management-and-Industrial-Hygiene/book-series/ CRCWRPLSAFRK

The Cause, Effect, and Control of Accidental Loss

Second Edition

Ron C. McKinnon

CRC Press
Taylor & Francis Group
Boca Raton London New York

CRC Press is an imprint of the
Taylor & Francis Group, an **informa** business

Designed cover image: © Shutterstock

Second edition published 2023
by CRC Press
6000 Broken Sound Parkway NW, Suite 300, Boca Raton, FL 33487-2742

and by CRC Press
4 Park Square, Milton Park, Abingdon, Oxon, OX14 4RN

CRC Press is an imprint of Taylor & Francis Group, LLC

© 2023 Ron C. McKinnon

First edition published by CRC Press 2000

Library of Congress Cataloging-in-Publication Data
Names: McKinnon, Ron C., author.
Title: The cause, effect and control of accidental loss / Ron C. McKinnon.
Other titles: Cause, effect, and control of accidental loss with accident investigation kit
Description: Second edition. | Boca Raton : Taylor & Francis, 2023. |
Series: Workplace safety, risk management, and industrial hygiene | Originally published: The cause, effect and control of accidental loss. Boca Raton : Lewis Publishers, c2000. | Includes bibliographical references and index. |
Identifiers: LCCN 2022052259 (print) | LCCN 2022052260 (ebook) | ISBN 9781032409054 (hardback) | ISBN 9781032474021 (paperback) | ISBN 9781003385943 (ebook) | ISBN 9781032474038 (ebook other)
Classification: LCC T55 .M385 2023 (print) | LCC T55 (ebook) | DDC 363.11—dc23/eng/20221108
LC record available at https://lccn.loc.gov/2022052259
LC ebook record available at https://lccn.loc.gov/2022052260

ISBN: 9781032409054 (hbk)
ISBN: 9781032474021 (pbk)
ISBN: 9781003385943 (ebk)
ISBN: 9781032474038 (eBook+)

DOI: 10.1201/9781003385943

Typeset in Times
by codeMantra

Contents

PART ONE   *Accidental Loss – The Cause*

PART TWO   *Accidental Loss – The Effect*

PART THREE   Accidental Loss – The Control

Preface

Accidents, sometimes termed unplanned, uncontrolled, and therefore undesired events, result in some form of a loss. The loss could be injury or illness, damage to property or equipment or some form of interruption to the process.

Accidents are caused. They have identifiable *causes*; they have an *effect* and can to a certain degree be *controlled*. Acts of nature are beyond our normal control, but only result in a small percentage of accidents. The majority of undesired events termed accidents can therefore be prevented.

BLAMING THE WORKER

Accidents have plagued humanity from the beginning of time. At one stage during our evolution, accidents were accepted as being merely part of life. Archaic safety philosophies of the 1930s found a means to blame workers for accidents and leave the employer blame free. Being convenient, this philosophy is still believed by many today. The fact that humans make mistakes and err is a no brainer. Blaming workers for accidents does not fix the problem.

Early safety thinking was that by eliminating the unsafe acts (high-risk behavior) of workers, the majority of accidents would be prevented. Unfortunately, that has proved to be more difficult than expected, and almost impossible. To solve a problem, the real or root cause must first be found and rectified.

ROOT CAUSES

Modern safety thinking is that there are deep-seated reasons why employees commit unsafe acts (high-risk) and why unsafe (high-risk) workplace conditions exist and are tolerated. These reasons can be determined by accident root cause analysis and can be eliminated or controlled to a tolerable level of risk. The loss causation theory presented in this book is an effort to redirect an organization's energies from trying to fix the worker, to rather fixing the workplace. Only by understanding the anatomy of an accident can we prevent its occurrence.

LUCK FACTORS

The introduction of the three Luck Factors into the accident sequence may seem controversial, but when one examines their influence on the outcomes of undesired events in the accident sequence, it may change the way we look at safety today.

SITUATION

Even today, safety is measured by the number and severity of injuries. Many are still convinced that the majority of accidents are caused by high-risk behavior of workers. The internationally accepted measure of safety, the disabling or lost-time injury rate, is an unreliable measure of safety performance, as accidental losses are the end result of numerous luck factors within the accident sequence.

PURPOSE

The purpose of this book is to examine the factors that come together and result in an undesired event which results in accidental loss. The loss may be in the form of personal injury, property damage, or business interruption. The same sequence of events that result in high potential near-miss incidents will also be examined. The book will analyze the entire process that constitutes the chain reaction of accidents and near-miss incidents.

FRAMEWORK

The framework for the book is the Cause, Effect, and Control of Accidental Loss (CECAL) sequence, or accident domino sequence, depicting the sequence of events that culminate in accidental loss. This 15-domino loss-causation model was originally proposed by McKinnon in the book *Contemporary Issues in Strategic Management,* edited by Elsabie Smith, Nicholas I. Morgan, et al. The proposed CECAL accident sequence was published on page 283.

SAFETY PIONEERS

The motivation for the modification, update, and reexamination of the components of an accident was inspired by writings of safety pioneers such as H. W Heinrich, James Tye, H. J. Matthysen, Frank E. Bird, Dan Petersen, and others. The work of these safety pioneers has been expanded on to compile the CECAL domino sequence. Ted S. Ferry and Weaver (1976) summarized the objective of this book by saying:

> The domino theory is durable. First presented by Heinrich about 1929, it has been updated by several persons. In 1976 Bird used a newer sequence with five dominoes identified as lack of control, basic causes, immediate causes, the incident, and injury to people and / or damage to property.
>
> *(p. 143)*

Even today Heinrich's domino sequence is still referred to and used as part of safety, risk management, and occupational hygiene-training courses. As with all theories, safety must be kept evergreen to keep pace with changing technology, people, and norms in general.

Ferry also mentioned the seven-domino sequence cited by Marcum. He said that in Heinrich's fifth book, Dan Petersen continued the tradition.

Another assumption that has not changed over the years is that the majority of accidents are caused by the high-risk behaviors (unsafe acts) of people. Heinrich's loss causation model and his theorems on safety were so radical to the safety industry at the time that readers immediately accepted his statement that 88% of accidents were caused by the high-risk behaviors (unsafe acts) of people and this is still believed, incorrectly, today.

Fred A. Manuele PE, CSP, (1997) describes this scenario in a nutshell:

> Heinrich's causation model has prominently been used by safety practitioners. Other causation models are extensions of it. However, the wrong advice is given when such models and incident analysis systems focus primarily on characteristics of the individual; unsafe acts being a prime cause of incidents; and measures devised to correct "man failure," mainly to affect an individual's behavior. Heinrich also wrote, "...a total of 88% of all industrial accidents ... are caused primarily by the unsafe acts of persons." Those who continued to promote the idea that 88 or 90 or 92% of all industrial accidents are caused primarily by the unsafe acts of persons, do the world a disservice.
>
> *(p. 31)*

Dan Petersen (1996) is quoted often in this book and this quotation summarizes the purpose of the CECAL study:

> Our present framework of thinking in safety should be examined, and perhaps challenged more than it is. Much of what we do today is based on principles developed long ago. It may now be time to re-examine those principles and look at some newer principles that have come upon the safety scene.
>
> *(p. 7)*

IMPORTANCE

The CECAL loss causation analysis of an accident is of vital importance to the safety management profession. It calls for a different way of looking at, measuring, and promoting the prevention of occupational injuries, diseases, and damage. This book clearly demonstrates that traditional forms of safety performance measurement, accident prevention, and the near disregard of near-miss incidents have to change before the accident toll can be reduced.

Acknowledgments

With so much information and work that has culminated in this document, numerous people need to be thanked. Thanks go to Bunny Matthysen, previous managing director of NOSA and inductee into the Safety Hall of Fame, Frank E. Bird Jr., founding director of the International Loss Control Institute, James Tye of the British Safety Council, Carel Labuschagne, managing director of International Risk Control Africa, John Bone, retired managing director of NOSA, the "NOSA boykies" (five-star representatives) of BHP Copper North America, San Manuel Mine and Plant. Bryan Wollam deserves recognition for the inspiration. Finally, my wife, Maureen McKinnon, who spent numerous weeks typing and editing this manuscript warrants my deep gratitude.

The contents of this document are dedicated to the thousands of people who have died as a result of occupational injuries and diseases, and to the millions who have been and are injured every year in industries and mines around the world.

About the Author

Ron C. McKinnon, CSP (1999–2016), is an internationally experienced and acknowledged safety professional, consultant, author, motivator, and presenter. He has been extensively involved in safety research concerning the cause, effect, and control of accidental loss, near-miss incident reporting, accident investigation, safety promotion, and the implementation of health and safety management systems for the last 46 years.

The author received a National Diploma in Technical Teaching from the Pretoria College for Advanced Technical Education, a Diploma in Safety Management from the Technikon SA, South Africa, and a Management Development Diploma (MDP) from the University of South Africa, in Pretoria. He received a Master's degree in Health and Safety Engineering from the Columbia Southern University.

From 1973 to 1994, Ron C. McKinnon worked at the National Occupational Safety Association of South Africa (NOSA), in various capacities, including General Manager of Operations and then General Manager Marketing. He is experienced in the implementation of health and safety management systems (SMS), auditing, near miss-incident and accident investigation, and safety culture change interventions.

From 1995 to 1999, Ron C. McKinnon was safety consultant and safety advisor to Magma Copper and BHP Copper North America, respectively. In 2001, Ron spent two years in Zambia introducing world's best safety practices to the copper mining industry. After leaving Zambia, he was recruited to assist in the implementation of a world's best class safety management system at ALBA in the Kingdom of Bahrain.

After spending two years in Hawaii at the Gemini Observatory, he returned to South Africa. Thereafter he contracted as the Principal Health and Safety Consultant to Saudi Electricity Company (SEC), Riyadh, Saudi Arabia, to implement a world's best practice safety management system, throughout its operations across the Kingdom involving 33,000 employees, 27,000 contractors, nine consultants, and 70 Safety Engineers.

Ron C. McKinnon is the author of *"Cause, Effect, and Control of Accidental Loss"* (2000), *"Safety Management, Near Miss Identification, Recognition and Investigation"* (2012), *"Changing the Workplace Safety Culture"* (2014), *Risk-based, Management-led, Audit-driven Safety Management Systems"* (2016), *"The Design, Implementation and Audit of Occupational Health and Safety Management Systems"* (2020), and *"A Practical Guide to Effective Accident Investigation"* (2022), all published by CRC Press, Taylor & Francis Group, Boca Raton, USA. He is also the author of *"Changing Safety's Paradigms,"* published in 2007 by Government Institutes, USA, and the second edition published in 2018.

Ron C. McKinnon is a retired professional member of the ASSP (American Society of Safety Professionals) and an honorary member of the Institute of Safety Management (South Africa). He is currently a health and safety management system consultant, safety culture change agent, motivator, and trainer. He is often a keynote speaker at health and safety conferences and consults to international organizations.

Data Collection

Data collection for the proposal of the CECAL accident causation sequence has been done by means of:

- research
- case studies
- incident recall sessions
- real life examples
- numerous information sources

RESEARCH

The research that has culminated in the CECAL sequence has included more than 46 years of international safety experience by the author as well as seven years in the electrical contracting industry. During that time, numerous safety surveys, safety gradings, health and safety management system (SMS) audits, accident investigations, safety training presentations, and consultations were conducted. Hundreds of safety interviews were held and the SMS in existence in numerous major organizations were examined, inspected, and audited.

The practical experience gained by the author in countries such as South Africa, Zambia, Botswana, Namibia, Canada, Hong Kong, Sweden, Chile, Peru, Australia, Bahrain, Saudi Arabia, United Kingdom, and the United States has culminated in the CECAL theory. Reference is often made to the book *Five Star Safety, an Introduction*, by the author. This is a 500-page manuscript, which was written in 1995. Although not yet published, this book has been used as a valuable source for the CECAL book.

CASE STUDIES

Selective case studies from the author's past experience have been used to emphasize certain aspects of the CECAL theory. Past events, both national and international, have been referred to and the respective sources quoted.

Numerous case studies have been referred to and these include fatal accidents, accidents that have resulted in injury, and those that have been termed near-miss incidents. To protect the persons involved in these case studies, names, dates, places, and other details have been omitted to avoid any involvement, litigation, or embarrassment. The case studies have been used only to emphasize the relationship between real life situations and the theories proposed by CECAL.

REAL LIFE EXAMPLES

As with the case studies, actual life examples have been used to expand on ideas and concepts in the book. The Luck Factors 1, 2, and 3 are certainly radical changes to traditional safety thinking and the author has taken the liberty of using these actual life examples to emphasize the existence of these luck factors.

Actual near-miss incident reporting systems have been referred to. Once again, the names of the organizations, individuals involved, and other details have been modified for obvious reasons. Some of the examples given have been extracted directly from organizations' health and safety management systems currently in operation, as well as from statistics obtained from them.

SOURCES

Numerous sources have been used to research this document and they include books, safety and health publications, websites, and periodicals. More than 50 sources are quoted in the attached reference list. Personal experience has been referred to in a number of instances.

Part 1

Accidental Loss – The Cause

1 Hazard Identification and Risk Assessment

THE CAUSE, EFFECT, AND CONTROL OF ACCIDENTAL LOSS ACCIDENT SEQUENCE

The Cause, Effect, and Control of Accidental Loss (CECAL) domino sequence proposes that all forms of accidental loss are triggered by the failure to identify the hazards, to analyze and evaluate the risk, and institute applicable risk-control measures in the form of a structured health and safety management system (SMS).

This in turn leads to an absence of or weaknesses in the SMS, which gives rise to workplace and personal factors, commonly referred to as the root causes of accidents.

These root causes then lead to high-risk behaviors being committed, and in turn allow high-risk workplace conditions to exist. These are the immediate causes that lead to an exposure, impact, or exchange of energy.

Once this situation exists, fortuity in the form of Luck Factor 1 determines whether there will be an exposure, impact, or contact with a source of energy that will cause a loss. No contact with a source of energy results in a near-miss event, close call, or as is defined in this book, a *near-miss incident*.

Should there be an exposure, impact, or contact with a source of energy, fortuity in the form of Luck Factor 2 then determines the outcome of this exposure, impact, or exchange of energy. The outcome could be illness or disease, injury, property damage, or business interruption or a combination of two, three, or all four.

If the exposure, impact, or exchange of energy causes personal injury, Luck Factor 3 then determines the severity of the injury.

The 15th domino in the cause-and-effect sequence depicts the costs that are incurred as a result of any undesired event, which results in a loss.

The last domino (domino 16) is the stabilizing domino which symbolizes a structured SMS, which in turn stabilizes the hazard identification and risk assessment domino, the weak SMS domino, and the root causes domino. This indicates that the sequence of an accident can be stopped by implementing and maintaining a structured SMS which prevents the sequence from being triggered.

In summary, the failure to identify the hazards, analyze and evaluate the risks, and set up control measures in the form of a structured SMS triggers off a chain of events that lead to accidental loss (Figure 1.1).

CONTROL

Failure to identify the hazards and assess the risks brought about by the business processes leaves the health and safety elements, processes, procedures, and items that need to be controlled, unidentified.

DOI: 10.1201/9781003385943-2

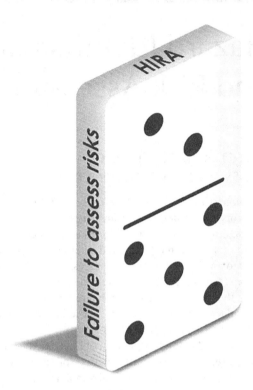

FIGURE 1.1   The first domino that triggers off the domino accident sequence represents the failure to identify hazards and assess the risk (HIRA).

As a result of poor controls in the form of an absent, weak, or inadequate SMS, personal and job factors (accident root causes) arise that lead to the creation of high-risk work conditions and high-risk behaviors (accident immediate causes). These eventually lead to the exposure, impact, or contact with a source of energy which causes property damage, injury, or loss. This chain of events culminates in the last domino in the sequence, which is financial losses.

> Effective health and safety management is not 'common sense' but is based on a common understanding of risks and how to control them brought about through good management.
>
> *Health and Safety Executive (HSE), Great Britain (1976)*

Hazard identification and risk assessment (HIRA) is a method that is predictive and can indicate potential for accidental loss. With this knowledge, an organization is then able to set up the necessary management controls within their SMS to prevent these risks resulting in losses such as injuries, property damage, business interruptions, and environmental pollution.

Many health and safety management efforts focus on the consequence of loss and not the control. Effective hazard identification and risk assessment is proactive, predictive safety in the finest form. In risk assessment the keywords are, "It's not what happened, but what *could have* happened."

Vernon L. Grose (1987) talks about the risk management approach as follows:

> The systems approach is a means of looking at even a very large problem in its entirety. You can start by calling whatever you are trying to manage – company, agency, product, family, organization, project, or farm – a system. Then you mark off its boundaries and define its inputs and outputs.
>
> *(p. 11)*

Grose says that this allows one to attack all the risks at one time.

HAZARD

A *hazard* can be defined as: "a situation which has potential for injury, damage to property, harm to the environment or a combination of all three." A hazard is a source of potential harm and high-risk behaviors, and high-risk work conditions are examples of hazards.

Some of the hazard classifications are:

- Safety – slipping and tripping hazards, poor machine guarding, equipment malfunctions or breakdowns.
- Biological – bacteria, viruses, insects, plants, birds, animals, and humans, etc.
- Chemical – depends on the physical, chemical, and toxic properties of the chemical.
- Ergonomic – repetitive movements, improper workstation setup, etc.
- Physical – radiation, magnetic fields, pressure extremes (high pressure or vacuum), noise, etc.
- Psychosocial – stress, violence, etc.

Hierarchy of Control of Hazards

There should be multiple layers of controls protecting employees from hazards and associated risks. The types of controls could include:

- Elimination – Which would mean changing the way the work should be done.
- Engineering controls – Such as the re-design of a worksite, equipment modification, and tool re-design.
- Administrative controls – Introducing a change of work methods and re-scheduling of work.
- Health and safety management system (SMS) controls – Such as health and safety policies and standards, work procedures, training, and personal protective equipment (PPE) (SMS elements, practices, procedures, actions, and programs).

RISK

A *risk* can be defined as "any probability or chance of loss." It is the likelihood of an undesired event occurring at a certain time under certain circumstances. The two major types of risks are *speculative* risks, where there is the possibility of both gain and loss, and *pure* risk, which offers only the prospect of loss.

Risk Management

Risk management combines the safety management functions of safety planning, organizing, leading, and controlling of the activities of a business to minimize the adverse effects of accidental losses produced by the risks arising from the organization.

Physical Risk Management

Physical risk management and financial risk management are the two main components of the science of risk management, and the best indicator for safety controls is physical risk management.

Physical risk management consists of identifying the hazards, assessing the risks, evaluating them, and introducing the necessary controls to reduce the probability of these risks manifesting in loss.

Risk Assessment

Risks cannot be properly managed until they have been assessed. The process of risk assessment can be defined as "the evaluation and quantification of the likelihood of undesired events and the likelihood of injury and damage that could be caused by the risks." It also involves an estimation of the results of undesired events.

One of the biggest benefits of risk assessment is that via risk evaluation; it will indicate where the greatest gains can be made with the least amount of effort, and which activities should be given priority. The safety management system now has a prioritization system based on sound risk assessment practices.

Components

Risk assessment and control has four major components:

1. hazard identification
2. risk analysis
3. risk evaluation
4. risk control

Once the first three phases of risk assessment are completed, risk control is then implemented. Risk control can only be instituted once all hazards have been identified and all risks quantified and evaluated. There is a substantial difference between a systematic approach to workplace health and safety and a behavioral-based safety approach.

The systems approach takes an objective and unbiased view of the workplace by:

1. identifying hazards
2. estimating the level of risk for each hazard
3. controlling hazards according to the hierarchy

According to Grose (1987):

> Its identification process limits a risk prevention program. If a risk is not first identified, it can never be evaluated or controlled.
>
> *(p. 13)*

HAZARD IDENTIFICATION

The first step of a risk assessment is the identification of all possible hazards. A hazard is a situation that has potential for injury or damage to property or the environment. It is a situation or action that has potential for loss.

There are numerous hazard identification methods and techniques. The two main techniques are the *comparative* and the *fundamental* methods.

FUNDAMENTAL TECHNIQUES

Fundamental techniques include Hazard and Operability studies (HAZOP); Failure Mode and Effect Analysis (FMEA); Failure Mode, Effect, and Critical Analysis (FMECA); "What if" analysis, checklists, and other techniques such as:

- hazard surveys
- hazard indices
- accident reports
- near-miss incident reports
- critical task identification
- safety management system (SMS) audits

HAZOP (HAZARD AND OPERABILITY STUDY)

A HAZOP involves taking a component of a system (e.g., a valve, or an element of procedure) and stressing it beyond its designing tension and normal operation. In conducting the HAZOP, components and systems of a plant are subject to guide-words applied to relevant physical properties such as temperature, flow, and pressure. During this process, a potential cause of a deviation or failure is sought, and a consequence defined. If the consequence is undesirable, then the hazard must be addressed by removal, mitigation, or control (Figure 1.2).

Item No	Guide word	Possible Causes	Consequences	Action	Person Responsible
1	Position	Moved unintentionally	Bearing no longer aligned	Erect a bump protection device	James Hargrieves

FIGURE 1.2 A HAZOP report sheet showing the guidewords used, possible causes, consequences, and mitigating action with responsibility delegated.

Item	Effect
The main generator can fail	No electricity generated, no light & heat
The transformer could fail	No power transmission

FIGURE 1.3 Hazard identification: Failure Mode and Effect Analysis (FMEA).

FMEA (FAILURE MODE AND EFFECT ANALYSIS)

The Failure Mode and Effect Analysis (FMEA) is another method of identification of hazards. This method is used for identifying possible failures in the system and resulting consequences. FMEA asks the question, "What system could fail and what would the effect be?"

An example of a FMEA method of hazard identification is given in Figure 1.3, which shows extracts of a risk assessment done in a power-generating unit. The FMEA exercise identified the main systems which could fail within the department and the consequences because of main system failures. A FMEA asks the question, "What can fail and what will the effect be?"

FMECA (FAILURE MODE, EFFECT, AND CRITICALITY ANALYSIS)

The Failure Mode, Effect, and Criticality Analysis (FMECA) is a hazard identification method that goes into more depth than the FMEA. The FMECA method examines each component of a system for criticality and identifies the effect on the entire system upon failure of specific components. This helps focus on critical components within a unit (Figure 1.4).

COMPARATIVE TECHNIQUES

Comparative techniques use checklists based on industry standards or existing codes of practice. They could involve comparing the plant in question with similar plants.

ETA (EVENT TREE ANALYSIS)

The Event Tree Analysis (ETA) is a predictive method of determining the cause and effect of events. The ETA starts with the event and deduces by means of Boolean logic what factors could contribute to the event.

Component	Effect
The brushes could short out	No power would reach the exciter coils
The insulator clip could break	Transformer failure

FIGURE 1.4 An example of a FMECA analysis. The questions asked are, "What components are critical to the process and what will the effect be if they fail?"

FTA (FAULT TREE ANALYSIS)

Fault Tree Analysis (FTA) is deductive as it deduces the events and sub-events that lead to the main event using the same method as ETA.

PAST ACCIDENTS AND NEAR-MISS INCIDENTS

A useful method of predicting future hazards is to review past injury and property damage causing accidents as well as high-potential near-miss incidents. By studying past loss-producing events, a pattern can be derived that would indicate certain recurring and inherent hazards within the business.

Near-miss incidents, or events, which under slightly different circumstances could have resulted in a loss, are perhaps the best indicators of the presence of hazards arising from the risks of the business.

Case Study

The case study on a near-miss incident in Figure 1.5 clearly indicated ten major hazards that existed that *could have* resulted in accidental losses worth more than $50,000. These hazards included poor leadership, lack of management training, no emergency preparedness, and little or weak employee training. These hazards

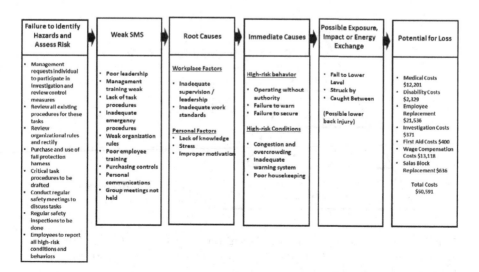

FIGURE 1.5 Loss causation analysis of a near-miss incident.

extended to the purchasing and engineering controls. This example clearly indicates that near-miss incidents can often highlight hazards that can then be controlled before the loss occurs.

INCIDENT RECALL

Formal and informal incident recall sessions are imperative if hazard identification is to be done thoroughly. Incident recall is a method whereby employees recall past near-miss incidents that under slightly different circumstances could have resulted in accidental loss. The losses could cause injury to people, property damage, or interruption to the work process.

Near-miss incidents are also vital in hazard identification. Near-miss incident reporting systems have often failed in the past yet can be very successful if they are kept anonymous with no repercussion after reporting an incident.

Sample

Over a 20-year period more than 500 employees were interviewed and asked why near miss-incidents were not reported. The two main reasons that inhibited the reporting were:

1. Fear of disciplinary action should they report an event that might have been caused by a high-risk act on their own or a colleague's part.
2. Whenever they report a near-miss incident they are confronted by, "You saw it, so why didn't you fix it?"

In referring to the accident ratio (Bird and Germain, 1992), and updated accident ratios, it is estimated that nearly 1,000 near miss-incidents could occur before a serious injury happens (Figure 1.6).

FIGURE 1.6 The accident ratio conclusion. (From McKinnon, Ron C. 2012. *Safety Management, Near Miss Identification, Recognition, and Investigation*. Model 2.4. Boca Raton, FL: Taylor & Francis. With permission.)

The accident ratio conclusion proposes that for one serious injury experienced at a workplace, there are *some* minor injuries, *more* property damage accidents and *plenty* of near-miss incidents with the potential for loss.

Example

One workplace received on average some 200 near-miss incidents reported each month. Most of them indicated hazards with potential to cause accidental loss. The reporting form also included a risk matrix so that the reporter had already conducted a mini risk assessment of the hazard during the reporting.

CRITICAL TASK IDENTIFICATION

A systematic method of listing work tasks and applying a critical task hazard analysis will highlight hazards or hazardous steps within certain tasks. This will enable written safe work procedures to be written for these critical tasks. These will help guide the employee to avoid risks when carrying out the critical tasks. Past injury and loss experience as well as probabilities are examined. Near-miss incidents are also considered in the critical task evaluation.

HEALTH AND SAFETY MANAGEMENT SYSTEM AUDITS

Structured audits of the existing SMS also indicate what elements and items are not functioning and which pose a hazard. These audits point out a company's SMS strengths and weaknesses and indicate which areas may pose more hazards.

SAFETY INSPECTIONS

A thorough safety inspection guided by a hazard control checklist is one of the basic and best methods of identifying physical hazards, high-risk conditions, as well as high-risk behaviors. Once identified, the hazards should be ranked by using a simple system such as the A, B, and C classification. An A-class hazard has the potential to cause death, major injury, or extensive business interruption, a B-class hazard has the potential to cause serious non-permanent injury and minor disruption of the business, and a C-class hazard has potential to cause minor injury and non-disruptive business interruption.

RISK ANALYSIS

Once hazards have been identified and prioritized in a hazard-ranking exercise (which follows the hazard identification process), the second step in risk assessment is then the analysis of the risks.

Risk analysis is "the calculation and quantification of the probability, the severity, and frequency of an undesired event occurring because of the risk." It is a systematic measurement of the degree of danger in an operation and is the product of the *probability* of occurrence, the resultant *severity* of the outcomes, and *frequency* of the undesired event.

PURPOSE

The purpose of the risk analysis is to reduce the uncertainty of a potential accident situation and to provide a framework to look at all eventualities. A risk analysis is a risk quantification method that looks not at *what happened* in the past but what *could happen* in the future. It is a method of identifying accidents that have not yet occurred.

RISK SCORE

The risk score is the product of the *probability* of the event's happening, the *severity* of the consequences, and the *frequency* of exposure of the event.

- The *probability* asks the question, "What are the chances of the event's happening?"
- The *severity* asks the question, "If it happens, how bad will it be?"
- The *frequency* asks, "If it happens, how often can it occur and how many people are exposed?"

Numbers are allocated to the various degrees of probability, severity, and frequency and the product gives a risk score.

Probability, Severity, and Frequency

The probability scale used in this example risk assessment (Figure 1.7) ranks the probability of damage as a (1) and a fatality as (6). The severity is ranked on a scale from (1) for a minor injury or damage, (4) for multiple disabling injuries or major damage, and (6) for multiple fatalities and/or permanent loss of structure. The

Risk Analysis						
Probability ➡	Damage 1	Fire 2	Explosion 3	Injury 4	Severe Injury 5	Fatality 6
	X	X	X	X	X	X
Severity ➡	Minor Injury/Damage 1	Serious Injury/ Damage 2	Permanent Disability/ Disruptive Damage 3	Multiple Disability/ Major Damage 4	Fatality/ Catastrophic Damage 5	Multiple Fatalities/ Permanent Loss of Structure 6
	X	X	X	X	X	X
Frequency ➡	Anytime 6	Hourly 5	Daily 4	Weekly 3	Monthly 2	Yearly 1
EXAMPLE RISK 'A': RISK SCORE = PROBABILITY (3) X SEVERITY (3) X FREQUENCY (3) =27						

FIGURE 1.7   An example of a risk analysis using probability, severity, and frequency.

frequency is (6), meaning it could happen "any time," to (1) for less often than yearly. The highest probability with the worst severity and the most exposure would be 6 multiplied by 6 multiplied by 6, which would be the highest risk score.

In Figure 1.7, Risk A ranked 27 (Probability (3) Severity (3) frequency (3) = 3 × 3 × 3 = 27). Other risks, as examples, not shown on the table, ranked as follows:

- Risk B ranked 80 – Probability (5) Severity (4) frequency (4) = 5 × 4 × 4 = 80
- Risk C ranked 36 – Probability (4) Severity (3) frequency (3) = 4 × 3 × 3 = 36
- Risk D ranked 60 – Probability (3) Severity (4) frequency (5) = 3 × 4 × 5 = 60
- Risk E ranked 2 – Probability (1) Severity (2) frequency (1) = 1 × 2 × 1 = 2
- Risk F ranked 216 – Probability (6) Severity (6) frequency (6) = 6 × 6 × 6 = 216

The risk analysis process has clearly indicated the highest risk and offers an opportunity for risk reduction prioritization.

RISK MATRIX

The risk analysis methodology can also be applied by using a risk matrix (Figure 1.8).

A risk matrix is a block diagram with two axes: the loss potential *severity* and *probability* of occurrence.

The *Probability of Occurrence* scale is:

- Rare   (1)
- Unlikely   (2)
- Possible   (3)
- Likely   (4)
- Certain   (5)

The *Possible Severity* scale is:

- Damage   (1)
- Minor Injury (2)
- Serious Injury (3)
- Major Injury (4)
- Fatality   (5)

Should the probability of the event be ranked as 20–25, then a thorough investigation of the event is then initiated. Should the ranking be slightly less, in the 9–16 range, a formal inquiry is initiated, and should the risk be ranked low, 4–10, a report is submitted for action. Lower ranked risks in the 1–5 range should be rectified on the spot or determined to be as low as is reasonably practical.

RISK FREE?

For all walks of life, business, and industry to be absolutely risk free would be improbable as well as impracticable. In risk reduction, the objective should be to

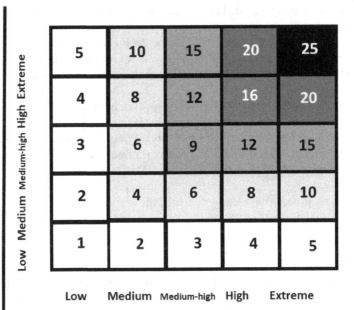

Potential Severity (How bad could it be?)

FIGURE 1.8 A risk matrix. (From McKinnon, Ron C. 2022. *A Practical Guide to Effective Accident Investigation*. Figure 1.1. Boca Raton, FL: Taylor & *Francis. With permission.*)

drive the risk As Low as is Reasonably Practicable (ALARP). The ALARP zone is where the risks have been reduced to where they can be tolerated, as the risk level is acceptable.

If the risks are kept in the ALARP zone it is regarded as normal business practice. Risks that extend above the ALARP zone could prove to be detrimental to the business as well as the employees and others working there.

RISK RANKING

Figure 1.9 shows the risk ranking matrix for six different risks and the risk score on the left-hand side ranging from 0 to 300. This model has been specifically orientated to highlight risks that have the potential for fatalities, and the fatality zone has been determined as being between 200 and 260, above that is a possibility of multiple fatalities. The risks that fall in the manageable zone, 80–180 need to be mitigated. The ALARP zone is between 0 and 60 and all risk reduction efforts are made to drive the risks out of the fatality zone and into the ALARP region. This is done with effective health and safety management control processes, procedures, and systems. In the example, Risk 5 falls in the fatality zone and Risk 4 is the only risk in the ALARP zone.

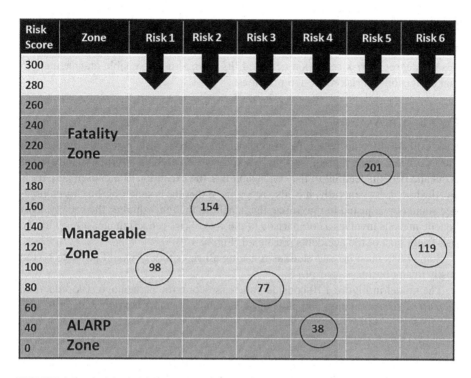

FIGURE 1.9 A risk ranking matrix, showing the fatality zone, the manageable zone, and the ALARP (As Low as is Reasonably Practicable) region.

RISK EVALUATION

The next step in the risk assessment process is risk evaluation. Grose (1987) explains risk evaluation as:

> Determining the accurate price of a preventive action requires both a skilled estimator and a consistent method of estimating labor costs, capital investments, and expendable outlays – whether one-time, periodic, or continuous.
>
> *(p. 15)*

He says that the price tag is critical and is what gets attention. Grose (1987) further explains the purpose of risk evaluation:

> To estimate the value of a preventive action logically, the risk control potential matrix forces consideration of *efficacy* (how much control will result), *feasibility* (how acceptable it will be), and *efficiency* (how much "bang for the buck" will result).
>
> *(p. 15)*

DEFINITION

Risk evaluation could be defined as: "the quantification of the risks coupled with an evaluation of the cost of degree of risk reduction and the resultant benefit derived from reducing the risk."

OBJECTIVE

The main objective of risk evaluation is to ensure that the cost of risk reduction justifies the degree of risk reduction achieved. Its main aim is to enable management to make decisions on risk reduction priorities.

COST BENEFIT ANALYSIS

Risk evaluation could be a type of cost benefit analysis. It is generally agreed that most decisions about human workplace activities are based on a form of balancing of costs and benefits leading to the conclusion that the execution of a chosen activity is worthwhile. Less generally, it is also recognized that the conduct of the chosen practice should be adjusted to maximize the benefit to the individual or the society. Cost benefit analysis involves a comparison of the cost of risk reduction measures with the risk-factor cost of the accidents prevented (Figure 1.10).

Some risks must be tolerated, as reducing all risks totally would prove to be too costly to any business enterprise.

The model in Figure 1.10 considers the risk score, the percentage risk reduction, the cost of risk mitigation and plots these variables on a justification scale of 0–100.

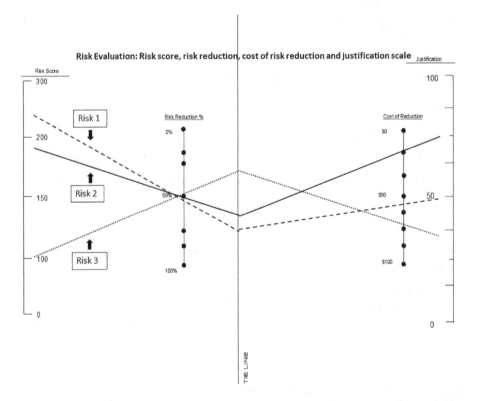

FIGURE 1.10   An example of a risk evaluation model showing risks score and cost of risk reduction versus percentage justification.

Join the risks score with the percentage reduction and plot a point on the tie line. A 50% risk reduction is recommended initially. Once the cost of a 50% risk reduction is determined, plot this figure on the cost scale. Join the point on the tie line with the point plotted on the cost and draw a line through to intersect the justification scale. Repeat this with the major risks and those that have the highest score on the justification scale should receive priority.

RISK PROFILE

A risk profile is a model charting the various risks in graph form according to the risk score. The justification scale is imposed on the same model. The first graph drawn in a solid line shows the risk score for each risk, and a second dashed line shows the justification for rectification as determined by the risk evaluation (cost benefit analysis).

Figure 1.11 shows a risk ranking profile for six major risks plotted in a solid line and the justification plotted by a dashed line. It is interesting to note that the risks with the highest score in some instances received the lowest justification. Cost of risk mitigation plays an important role in risk evaluation.

RISK CONTROL

The final step in the HIRA process is risk control. Risks cannot be managed until they have been assessed. Once the risks have been assessed and prioritized through the risk evaluation process, risk controls are now decided upon.

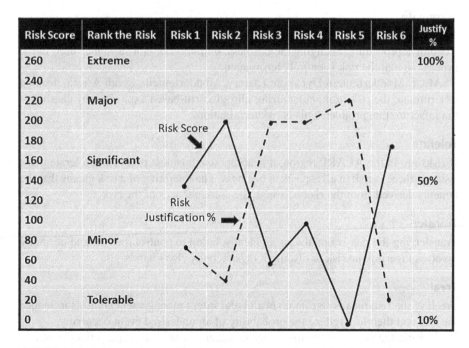

Risk Score	Rank the Risk	Risk 1	Risk 2	Risk 3	Risk 4	Risk 5	Risk 6	Justify %
260	Extreme							100%
240								
220	Major							
200								
180								
160	Significant							
140								50%
120								
100								
80	Minor							
60								
40								
20	Tolerable							
0								10%

FIGURE 1.11 Risk ranking profile and justification.

Once the risks have been assessed and the risk control method chosen, safety management control in the form of structured SMS standards, practices, and procedures are now applied.

OBJECTIVE

The objective of risk control is to minimize those risks that have been assessed and wherever possible to transfer them. Exposure to the risk is minimized and contingency plans drawn up to cope with the consequences should an event occur.

GOALS

The goals of risk control are to reduce the probability, severity, and frequency of undesired events occurring to a level as low as practicable.

RISK CONTROL METHODS

There are basically four ways to handle risks:

1. terminate the risk
2. tolerate the risk
3. transfer the risk
4. treat the risk

Terminate

The ideal method of risk mitigation is to terminate the risk entirely. This would mean stopping a procedure, changing the business, or disposing of a substance used in the process so that the risk is entirely terminated.

AECI, Modderfontein Dynamite Factory, Modderfontein, South Africa, decided to terminate the risk of manufacturing nitroglycerin-based explosives by ceasing to manufacture that product. This is risk termination.

Tolerate

If risks are in the ALARP region, it is acceptable business practice to tolerate these risks, as there is risk in all aspects of business. The tolerating of a risk means that the benefits derived from the risk outweigh the consequences of the risk.

Transfer

Transferring the risk is not always an ideal solution to control the risk and normally involves insuring the risk or placing it in somebody else's hands.

Treat

Treating the risk means setting up health and safety management controls to reduce the risk and therefore reduce the probability of an undesired event occurring.

Treating the risk involves the safety management principle of safety controlling. Safety controlling is the management function of identifying what must be done for safety, inspecting to verify completion of work, evaluating, and following up with safety action.

SAFETY MANAGEMENT CONTROL FUNCTION

The acronym IISSMECC is used to explain the control function where:

    I – Identify hazards and assess the risks.
    I – Identify the actions needed to reduce the risks.
    S – Set standards of accountability.
    S – Set standards of measurement.
    M – Measure against those standards.
    E – Evaluate conformances and non-conformances.
    C – Corrective action to be taken.
    C – Commendation for work well done.

The management function of *safety controlling* will be discussed in detail in Chapter 17.

REAL CAUSE

Failure to identify the hazards, assess, analyze, evaluate, and control the risks is the key event that triggers off the chain of events referred to as accidents (undesired events that result in undesired losses.)

Hazard identification, risk assessment, risk management, and risk control are the future of traditional health and safety as we know it and will lead to predicting where loss-occurring events may happen and enable us to prevent the accidents that have not yet occurred.

ACCIDENT ANALYSIS

In the analysis of a fatal accident, the initiating event that caused ten systems to be out of control was the failure to adequately identify the hazards and assess and control the risks.

Example

An investigation of a major equipment damage accident, using an analysis based on the *Cause, Effect, and Control of Accidental Loss* (CECAL) domino sequence, indicated that critical parts, critical tasks, and procedures had not been identified. No risk profile had been compiled for the process and no hazard identification or risk assessment had been carried out. This led to inadequate safety management system controls, which then triggered off the sequence of events resulting in the losses.

SUMMARY

Hazard identification and risk assessment forms a vital part of a good health and safety management system. It complements and builds upon the existing skills and techniques of line managers and safety professionals. Risk assessment is a management tool, which assesses performance, enables analysis, and creates goals and standards. Failure to identify the hazards and assess and manage the organization's risks triggers off the accident chain reaction (Figure 1.12).

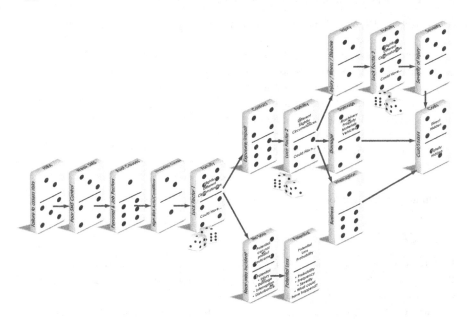

FIGURE 1.12   The Cause, Effect, and Control of Accidental Loss (CECAL) accident domino sequence, which shows how the failure to identify hazards and assess the risk gives rise to a weak or ineffective health and safety management system which in turn leads to the losses and subsequent costs.

2 Weak or Flawed Safety Management System (SMS)

WEAK CONTROL

If the hazards arising out of the business have not been identified, and their consequent risks have not been assessed and evaluated, they cannot be managed or controlled. This creates a lack of, or poor management control, in the form of a weak health and safety management system (SMS), which is depicted by the second domino in the chain of events leading to accidental loss.

HEALTH AND SAFETY MANAGEMENT SYSTEM

A health and safety management system (SMS) is defined as: "on-going activities and efforts directed to control accidental losses by implementing and monitoring critical health and safety elements of a SMS on an on-going basis." The monitoring includes the promotion, improvement, and auditing of these critical health and safety elements regularly. A SMS is the framework of policies, processes, and procedures used to ensure that an organization can fulfill all tasks required to achieve its objectives as defined in the health and safety policy statement. Implementing and maintaining a SMS is management control (Figure 2.1).

> ...there is a great tendency – human tendency – for management to rationalize after experiencing a human tragedy. It is always so much easier to find the "careless acts" on the part of an injured employee which precipitated the accident, but an enlightened management will not hesitate to look beyond the "unsafe act" on the part of an employee and to consider it as a symptom of lack of management control.
>
> *Lester A. Hudson (1995, p. 2)*

DIRECTION

It has often been said that if you do not know where you are going, any road will lead you there. Safety management is the same. Unless direction is set and actions are based on a thorough understanding of risks, all efforts to reduce accidents may be misdirected.

DOI: 10.1201/9781003385943-3

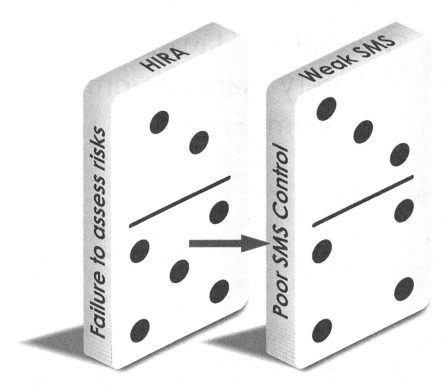

FIGURE 2.1 Failure to identify hazards and assess the risks leads to poor control in the form of weaknesses within the health and safety management system (SMS).

ALICE IN WONDERLAND?

In the classic book, *Alice's Adventures in Wonderland,* (Lewis Carroll), Alice and the Cheshire cat have a conversation:

> "Will you tell me please, which way I ought to go from here?" Alice asked the Cheshire cat. "That depends a good deal on where you want to get to" replied the Cheshire cat. "I don't care much," said Alice after some thought. "Then it doesn't matter which way you go" replied the Cheshire cat.
>
> *(p. 103)*

This brief conversation is the same as an organization not having set controls within a structured SMS to combat accidental losses. If you do not know where you are going, any road will lead you there.

THREE CONTROL OPPORTUNITIES

Management has opportunity to control at three stages within the cause-and-effect sequence.

Pre-contact Control: Pre-contact control is the management work done to prevent accidental loss before it occurs. It consists of all the activities of managing the business and setting up controls in the form of standards and systems within a structured

SMS. This will ensure that controls and guidelines are set in place to reduce the probability of the risks manifesting as accidental losses.

Contact Control: Contact control is where the undesired event has not been prevented, but the consequences of the exposure, impact, or exchange of energy have been minimized. Contact control does not prevent the accident per se but merely minimizes the result. Providing personal protective equipment and ensuring that workers wear it is an ideal example of contact control. Personal protective equipment does not prevent a brick falling from an overhead scaffold but does minimize the amount of energy exchanged between the brick and the person standing below, and consequently does exercise a degree of contact control.

Post-contact Control: Post-contact control includes the management activities that are conducted after there has been an exposure, impact, or exchange of energy. After a near-miss or loss-producing event, something is done to fix the problem. Examples of post-contact control are:

1. Accident investigation
2. Revision of procedure
3. Injury statistic compilation
4. Removing the hazard
5. Retraining employees
6. Hazard elimination
7. Modifying the process, etc.

SAFETY MANAGEMENT FUNCTIONS

A manager's four main functions, according to Louis A. Allen, are planning, organizing, leading, and controlling. These are also management's safety functions.

SAFETY PLANNING

Safety planning is what a manager does to predetermine the occurrence and consequence of accidents and to determine action that must be taken to prevent downgrading events occurring. This includes hazard identification and risk assessment (HIRA) and risk abatement.

SAFETY ORGANIZING

Safety organizing is the function a manager carries out to arrange safety work to be done most effectively by the right people. This entails allocating health and safety authority, responsibilities, and accountabilities to the right levels of management.

SAFETY LEADING

Safety leading is what a manager does to set an example and ensure that workers act and work in a safe manner. It also involves visible felt management leadership and the setting of an example for safety.

SAFETY CONTROLLING

Safety controlling is the management function of identifying what work must be done for health and safety, setting standards of measurement and accountability, inspecting to verify completion of work, evaluating, and following up with safety action.

Health and Safety Management System

Safety controlling is the process used in the developing, implementing, and maintaining of a risk-based, management-led, audit-driven SMS.

IISSMECC

According to *Questions and Answers*, NOSA (National Occupational Safety Association) (1998), there are six steps in the controlling function.

1. Identification of work situations that involve a high element of risk
2. Setting standards of performance and measurement
3. Setting standards of accountability
4. Measurement against the standards
5. Evaluation
6. Correction (p. 13)

Two further steps can be added to this control process. They are the identification of the risks of the business, which include conducting hazard identification and risk assessments, and introducing risk controls to impact on, and direct, the work that needs to be carried out to minimize risk. The last aspect would be the commendation for compliance to standards and recognition of satisfactory performance. The modified process would then be as follows:

I – Identify hazards and assess the risk.
I – Identify the work that needs to be done to mitigate and control the risks.
S – Set standards of measurement.
S – Set standards of accountability.
M – Measure conformance to standards by inspection.
E – Evaluate conformances and achievements.
C – Correct deviations from standards.
C – Commend compliance.

The IISSMECC process and how it can be applied to facilitate the implementation of a comprehensive SMS will be discussed fully in Chapter 17, *Management Control*.

AREAS OF WEAK HEALTH AND SAFETY CONTROL

A lack of safety management control can be determined by the following:

* An inadequate health and safety management system (SMS)
* SMS standards, processes, and procedures that are inadequate for the risk

- The SMS standards are not complied with
- Inadequate audit of the SMS which does not reveal weaknesses
- Weaknesses in the SMS
- The SMS has not been updated according to changes in risks
- Non-compliance with local health and safety laws and regulations

Indicators of an inadequate, non-existent, or weak SMS could be:

- Non-enforcement of SMS rules, processes, and procedures
- Hazard reporting and rectification process not functioning correctly
- Irregular holding of safety committee meetings
- Health and safety meetings not led by line managers
- Fewer near-miss incidents are being reported
- Accident investigations not completed and remedial measures not implemented
- Deterioration in workplace housekeeping
- Workplace and other safety inspections not carried out correctly
- Deterioration in the quality of health and safety training, etc.

Figure 2.2 shows the results of an audit of seven elements of an SMS carried out in Year 1 and Year 2. The scores show a drop in audit scores of most elements in Year 2 indicating that there is a weakness in the SMS.

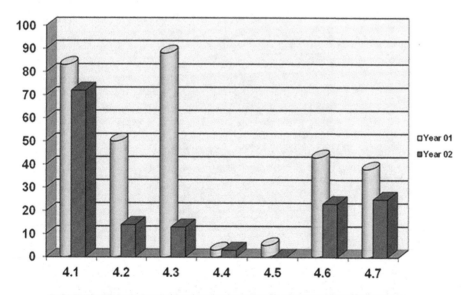

Element			Element	
4.1	Housekeeping		4.4	Near-miss reports
4.2	Inspections		4.5	Accident investigations
4.3	Hazard reports		4.6	Health and Safety Meetings
			4.7	Training attendance

FIGURE 2.2 The results of audit of seven SMS elements, showing the reduction in scores between Year 1 and Year 2.

INADEQUATE SMS

The SMS may be inadequate and does not address all the risks of the organization. Some systems, processes, or procedures may be bypassed or overlooked, or there may be non-compliance or disregard for the rules laid down in the standards. Reporting systems may have collapsed, and safety training has been neglected. There are a number of issues that could result in a weakness in the SMS and its processes and procedures.

Safety control eliminates the root causes of accidents by setting up management systems within the SMS, and by delegating safety responsibility. This system creates a work environment in which personal factors and job factors, the accident root causes, are reduced, consequently reducing high-risk behaviors and high-risk conditions.

As Dr. Mark A. Friend (1997) says, "Only members of the management team can create or change the environment (And it is, after all, their job to do so)" (p. 34).

CONCLUSION

Inadequate, weak, failing, or nonexistent controls in the form of a structured health and safety management system (SMS) gives rise to the root causes of accidents. This leads to high-risk situations, which then lead to an exposure, impact, or exchange of energy which results in a loss.

3 Root Causes of Accidental Loss

The lack of, or inadequate, management control in the form of weaknesses in the health and safety management system (SMS), caused by a failure to identify hazards, assess, and control risks, leads to the root causes of undesired events. These root causes lead to high-risk behaviors and conditions and are a vital link in the chain reaction that results in an undesired event with loss (Figure 3.1).

THE MANAGEMENT PRINCIPLE OF DEFINITION

The management principle of *definition* states that a logical and proper decision can only be made when the root or real problem is first identified. In safety, very seldom is the real problem in the form of accident root causes identified and treated. Most safety efforts are reactive, as they treat the symptoms and not the causes of accidents. It is easier to treat the symptom because the symptom is more readily identifiable, whereas in-depth delving and investigation, coupled with tenacity and risk taking, are needed to unearth the root causes. Prescription without diagnosis is malpractice.

As Dan Petersen (1998) wrote:

> By focusing on high-risk behaviors and conditions, safety managers end up dealing with accidents on a symptomatic level rather than a causal level.

(p. 39)

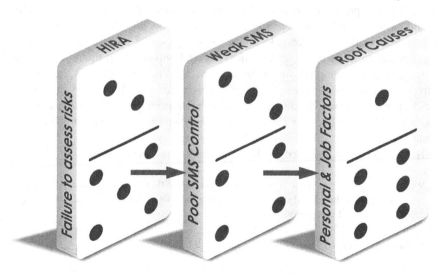

FIGURE 3.1 The third domino in the CECAL sequence represents the accident root causes.

DOI: 10.1201/9781003385943-4

EXAMPLE 1

After presenting an eight-hour safety management training course, one of the attend-
ees, a senior supervisor with many years of experience, approached me with a concerned look on his face. He started to tell me that the employees in his workshop had
a bad attitude toward safety. Startled by his statement, I then inquired what he meant
by a poor attitude to safety. He said that his employees constantly trashed the workshop floor by just throwing their litter and rubbish onto the floor with total disregard.
I was concerned by his problem and offered to visit the workshop with him.

Upon entering the workshop, I was struck by the total disorder that confronted me.
The lighting was poor, the ventilation systems malfunctioned, and the entire floor
was almost ankle deep in debris, dirt, superfluous material, filth, and grime. There
was no demarcation of work areas or stacking areas, no dirt bins or scrap containers
were available, and the floor was almost like a minefield with obstructions. Trip and
slip hazards were everywhere.

It was obvious that the root cause of their behavior was the work environment,
which encouraged them to have no respect for their surroundings. In this case, the
trashing of the floor was only a symptom of the root cause of the unsafe and unhy-
gienic work conditions in which they worked. There was clearly a lack of standards
for housekeeping, demarcation, stacking and storage, and removal of trash and waste.
Employee training was non-existent. All these weaknesses in the SMS were the
underlying or root cause of the workers' behavior. Their attitude towards safety was
influenced by their workplace environment.

FOCUS ON THE CAUSE NOT SYMPTOM

Often, we tend to focus on the obvious, which is the symptom of the issue and not the
cause. The symptoms are as a result of an underlying, hidden, or root cause. These
should be found and treated, as treating the symptoms will not solve the problem.
In health and safety, one tends to notice the symptoms of a problem before understanding its cause. If one reacts instinctively and without deep thought, one could
be mistaking that symptom for the problem itself. When this happens, any solution
implemented acts as a temporary and often misguided fix. This is also applicable in
accident investigations. As stated, the management principle of *definition* states that
a logical and proper decision can only be made when the *root* or real problem is first
identified.

Heinrich (1959) acknowledges the root causes of accidents as follows:

> While this is an important function, we now know that long-range improvements
> can best be made by identifying and correcting the root causes. These can be loosely
> grouped into three categories:
>
> • Management safety policy and decisions (Job factors)
> • Personal factors
> • Environmental factors (p. 35)

HIGH-RISK BEHAVIOR

Dr. Mark A. Friend (1997) wrote:

> Employees do what seems rational. When they take short cuts or refuse to wear personal protective equipment, it is because management has created an environment in which such actions are the rational way to respond.
>
> *(p. 34)*

This statement is one of the most important, controversial, yet true statements in safety management. Over the years, both safety practitioners and management have been convinced that accidents are largely due to the high-risk behaviors of workers. By focusing on the behaviors of workers they have tended to lose sight of the fact that the behaviors are merely a result of the climate that management creates, both within the work environment and by the work-safety ethics that it dictates (or should dictate). Workplace environments create safety culture.

ROOT CAUSES ARE DISEASES

Many refer to the root causes as the disease or real cause behind the symptoms. They motivate this by saying that this is because the immediate causes (the symptoms, the high-risk behaviors, and conditions) are usually quite apparent, it takes a bit of probing to get at root causes and to gain control of them. Root causes help explain why people perform high-risk practices.

TRAGIC ACCIDENT

A tragic accident occurred during a live stage performance when the stand-in lead actor left the stage, went backstage, and exited to the right where she fell off an unguarded balcony to the level 20 ft (6 m) below. Initial findings were that she was supposed to have turned to the left when exiting the stage. The lead role actor had fallen sick at short notice and the stand-in was brought in to take the lead role on this particular night.

Although she was seriously injured, the actor survived but was left with memory loss, hearing loss, and a deformed right arm. Despite her injuries she returned to the stage a year later.

Upon interviewing and investigating the accident, it was obvious that the immediate cause of the accident was the fact that she had turned the wrong way and there was an unguarded, poorly illuminated drop-off backstage.

During rehearsals she had been instructed to turn to the left when exiting the stage, but factors such as stress and the poor lighting backstage led to her taking the wrong turn and falling off the unguarded edge.

During the investigation it was noted that she had apparently failed to follow the instructions carefully. However, during a TV interview a few years later she was explaining the injuries to her hand and her arm and repeatedly indicated with her right arm while calling it her left arm. It was clear that one of the root causes of the accident was that she was under stress and also that she was apparently uncertain as to her left and right.

Left-Right Confusion

Apparently, *left-right confusion*, or *directional confusion*, as neuroscientists appropriately call the phenomenon, is quite common. More than 25% of college students and about 20% of college professors reported the problem in one study. In a 2020 study, 14.9% of the people surveyed said that they had difficulty distinguishing left from right.

Could this have been one of the root causes of the actor taking the wrong turn to exit backstage?

Categories of Root Causes

There are three categories of root causes:

- *Personal* (human or individual) *factors*, which relate to factors surrounding the worker's actions.
- *Job* (organizational, engineering, or workplace) *factors,* which relate to the organizational systems and the workplace.
- *Natural causes* (acts of nature) such as hurricanes, floods, earthquakes, and other events beyond our normal control.

Only personal and job factors will be discussed. Examples of personal factors are:

- inadequate physical or physiological capability
- inadequate mental (cognitive) or psychological capability
- physical or psychological stress
- mental or psychological stress
- lack of knowledge
- lack of skill
- improper motivation, etc.

The eight main job factor classifications are:

- inadequate leadership and/or supervision
- inadequate engineering
- inadequate purchasing
- inadequate maintenance
- inadequate tools and equipment
- inadequate work standards
- wear and tear
- abuse or misuse

Root Cause Analysis Example

During the investigation of a major property damage accident, it was discovered that the high-risk act of "failure to warn" had been committed. Further investigation revealed the reason for this "failure to warn" was that the wrong reading was given by the computer. This in turn did not warn the hoist operator as to the true position of the skips (elevators). There was no other verification or check and balance to warn the hoist operator that there was imminent danger.

Another high-risk behavior was that the operator operated the hoist at full speed (operating at unsafe speed). Following a root cause analysis, the reason for this was found to be an inadequate or non-existent procedure, which allowed the operator to operate at full speed before confirming the skip's position.

Root Cause Explained

Root cause analysis is explained by a quotation from a property damage accident which, under the heading of root cause analysis, reads as follows:

> Underlying or root causes are the real causes of the problem and give rise to the high-risk behaviors and conditions which are the symptoms. Preventative measures must address the root causes so that the symptoms do not re-occur. Each high-risk behavior and condition is examined and the root cause analysis done by asking the questions, why? why? why?

Accident Prevention

NOSA (1988) asks the question:

> To prevent the same accident being repeated would it be sufficient to eliminate the immediate causes? For example, the high-risk behavior and/or the high-risk condition?
>
> Answer: No! The high-risk behaviors and /or high-risk conditions (immediate causes) exist because of less obvious causes, (personal and job factors), which must be identified and eliminated.
>
> *(p. 20)*

An example is given as follows:

> Oil on the floor can cause accidents and injury. Identifying the oil as the immediate cause of the accident is correct but if we wipe up the oil, we are not treating the root cause of the oil being on the floor. If we investigate further, we may find a forklift truck leaking oil, which is the root cause. To eliminate this (root cause) the forklift truck should be repaired.
>
> *(p. 20)*

Example 2

During the investigation of an accident that resulted in injury, the technique of immediate and root cause analysis was applied. Seven high-risk conditions were identified as well as five high-risk behaviors. In sitting with the team and by using brainstorming techniques, a root cause analysis indicated that there were 13 root causes that had led to the high-risk conditions and high-risk behaviors.

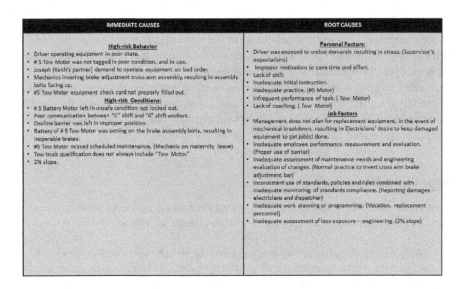

FIGURE 3.2   The immediate cause and the root cause analysis of an undesired event.

Example 3

Root causes create high-risk behaviors and high-risk conditions. These are the immediate causes of the contact that causes the loss. Figure 3.2 shows an accident investigation using root cause analysis. The root causes were found to be six job factors and seven personal factors. These created the conditions and actions that caused this loss-producing event.

ROOT CAUSES – PRECURSORS TO IMMEDIATE CAUSES OF ACCIDENTS

Root causes are the reasons that high-risk behaviors occur, and high-risk workplace conditions exist or are created. Root causes are the fundamental, underlying, system-related reasons why an accident occurs. They identify one or more correctable system failures.

Only once root causes have been identified, can meaningful management controls be put in place. Root causes are not easily recognizable, and a great deal of effort is needed to complete an effective root cause analysis.

In endeavoring to uncover the root causes of any loss causation situation, the immediate causes should first be identified via a thorough investigatory process. Only once the immediate causes are identified can root causes be derived.

Root Cause Analysis

A root cause analysis is a systematic method of determining the underlying (root) causes behind the obvious immediate accident causes of high-risk behavior and high-risk conditions. It establishes the reason why the act was committed and establishes why the high-risk condition existed. Accident root causes have also been called "the story behind the story."

QUESTIONING PROCESS

Root cause analysis is a structured questioning process that enables investigators to recognize and discuss the underlying beliefs, practices, and processes that result in accidents. Root causes are basic accident causal factors, which if corrected or removed may prevent recurrence of an accident.

Once the immediate causes are established, the root causes will indicate where the organization failed to identify the risks and put suitable health and safety systems in place. Implementation of applicable risk reduction controls to treat the root causes will prevent similar accidents occurring in the future. In the event of a high-potential near-miss incident, the exact same technique is followed.

THE 5-WHY METHOD OF ROOT CAUSE ANALYSIS

This is a simple but effective method of root cause analysis (problem solving) and involves asking the question "Why?" for each high-risk behavior and high-risk condition identified, at least five times, until a deep-seated reason for the failure is found.

In some cases, the answer to a "Why?" may be another immediate cause in the form of a high-risk behavior or condition. The investigator should then continue asking "Why?" for each immediate cause that surfaces, until all the root causes are found.

Evaluating and criticizing the answers before the process is completed will interrupt the questioning process and serve no purpose. The duplicated answers are a necessary part of the process of deriving the root causes.

AN EXAMPLE OF A ROOT CAUSE ANALYSIS

After an accident in which a worker suffered an eye injury, it was discovered that the machine shop bench grinder was in an unsafe condition. It had no fixed face shield, and the gap between the wheel and tool rest exceeded the 1/8th inch (3 mm) limit. The worker who fitted the wheel installed the wrong size wheel and removed the fixed face shield while he was removing the old wheel. He never replaced the shield.

High-risk Condition – The high-risk condition was an unguarded machine, as the gap between the wheel and the tool rest was too big, and the built-in face shield was missing.

The root cause analysis is as follows:

Why? – The face shield had been removed (Root Cause – Job Factor).

Why? – The worker removed it to fit the wheel (Root Cause – Personal Factor).

Why? – It was dirty, scratched, and hampered vision of the grinding job (Root Cause – Job Factor).

Why? – Grinders were never checked or maintained (Root Cause – Job Factor).

Why? – There were no regular inspections of bench grinders (Root Cause – Job Factor).

Why? – No one was responsible for grinder maintenance and inspection (Root Cause – Job Factor).

Why? – There was no maintenance schedule for these machines (Root Cause – Job Factor).

High-risk Behaviors – The face shield was not replaced, and the wrong size wheel was installed which left a gap between the wheel and tool rest that could not be adjusted.

Why? – The worker who fitted the wheel was inexperienced in replacing grinding wheels (Root Cause – Personal Factor).

Why? – He was working without supervision (Root Cause – Job Factor).

Why? – He forgot to replace the face shield (Root Cause – Personal Factor).

Why? – No one checked his work when he was finished mounting the wheel (Root Cause – Job Factor).

Why? – The risk of allowing inexperienced workers to maintain high-risk machines was not considered (Root Cause – Job Factor).

The root causes for the poor condition of the bench grinder were that the worker, who was unskilled in the task, was given the job of replacing the wheel. He was working without supervision, and no one checked on the completed task. He did not realize the importance of the face shield and neglected to replace it when he completed the job.

There were also no pre-use inspections or maintenance schedules for these machines, which led them to be operated while in a high-risk condition. No one had been assigned the responsibility of checking the machines after maintenance. Since they were not on the monthly workshop inspection check sheet, they were overlooked.

Safety Management System Standard

A Health and Safety Management System (SMS) standard for bench grinders should be introduced and should include the following:

- The correct wheel for the grinder.
- Regular dressing of the wheel.
- Fixed face shield guard condition.
- The regular adjustment to ensure a 1/8th inch (3 mm) gap between the tool rest and wheel.
- A notice to warn workers to wear protection to be erected at grinder.
- The wearing of personal eye protection when grinding.
- Regular inspections of grinders using a checklist.

FATAL ACCIDENT INVESTIGATION

During the investigation of a fatal accident, one of the high-risk behaviors committed was "failure to make secure." By applying root cause analysis, four personal factors were derived. They were (1) abuse or misuse, (2) stress, (3) lack of experience, and (4) lack of coaching. A further analysis was then done to uncover the reasons for the root causes. In this accident investigation, the reasons were:

Abuse or Misuse – there were three apparent reasons for this

1. Convenience – which allowed the job to proceed even though the scaffold was not secured.
2. Habit – site inspections revealed that several other scaffolds had also not been correctly secured.
3. Stress – five of the witnesses stated that they were under pressure to complete the work.

The example shows that there is always an underlying reason for a high-risk condition or a high-risk behavior. These reasons must be identified before the undesired event, rather than after. Conducting risk assessments and setting up controls to eliminate root causes is the best method of loss prevention. An important post-contact control method is accident investigation, which has the main objective of deriving and eliminating the root cause of the accident to prevent future recurrence. This is retrospective and reactive rather than proactive.

Example 4

During a serious accident investigation, the high-risk behavior of "ignoring rules" was identified as one of the immediate causes of the accident. This occurs frequently at workplaces and is often identified as the one and only reason that the event occurs. In this instance, the personal factors were sought, and at least eight reasons why the rules were ignored were uncovered. They included:

1. The rules had only been issued two weeks prior to the accident.
2. The operator in question was not in possession of the rules.
3. The rules had been blatantly and often ignored in the past with no consequence.
4. There were no inspections during the operation.
5. There was no control of contractors.
6. There was no follow-up to ensure compliance with rules.
7. The rules were found to be conflicting.
8. The instructor spent 90% of his time operating and only 10% in training (derived by personal interview).

During the same investigation it was discovered that the operator had operated at an unsafe speed (high-risk behavior). The analysis indicated seven root causes that indicated that this operator was in a hurry. They included:

1. It was nearing the end of the 12-hour shift.
2. The light was fading quickly (personal observation showed that it would have been dark within the next 15 minutes, making the task impossible).
3. The job had been delayed for some eight hours.
4. Improper motivation.
5. Extreme judgment, decision demands.
6. Confusing demands.
7. Inadequate instruction.

CAN OF WORMS

In attempting to derive root causes, the investigator may open a can of worms because traditionally, workers were blamed for committing high-risk behaviors. Certain organizations have made very generalized statements, which have unfortunately been believed and practiced by management, incorrectly, over a number of years. Root cause analysis will reveal the underlying cause of such behavior.

HIGH-RISK BEHAVIOR

In its training courses, Du Pont states that 96% of accidents are caused by the high-risk behaviors of people. On questioning this statement, the company says that people create high-risk conditions anyway, so that means people are guilty of creating high-risk conditions and of committing high-risk behaviors.

Unfortunately, this statement and argument is far too simplistic and far too generalized to enable us to clearly identify and analyze what causes accidents. This attitude also blinds us to the fact that there are reasons that people do what they do. It also blinds us to the fact that sometimes the work conditions are below accepted standards and may be hazardous.

According to the United Steel Workers International Union (2005):

> First, it is important to gain a better understanding of the role that DuPont Safety Training Observation Program (STOP), the company's behavioral-based safety program, plays in DuPont's approach to safety. STOP is grounded in the theory that almost all injuries are caused by worker unsafe acts and neglects many elements included in the National Safety Council's Hierarchy of Controls.
>
> *Assets-USW Website (2022)*

Investigating why high-risk behaviors are committed results in questioning the supervision and management control that existed (or did not exist) at the time of the event. This is what opens the first can of worms.

FOUR-FINGER RULE

The four-finger rule is the best method of explaining how root causes form an important part of the loss causation process. Immediately upon occurrence of an accident the tendency is to point a finger at the injured person and say, "He messed up," or, "Employee failed to …" While pointing a finger, it is interesting to note the position of the other three fingers. They are pointing directly back to management, asking, "Were we not responsible for this person's action? Have we set the standards, have we monitored the standards, is our behavior beyond reproach?" Answers to these questions may very well reflect that the root causes are poor or inadequate management control.

ANOTHER CAN OF WORMS

This leads to another can of worms being opened. Most safety professionals are hesitant to embark on this risky endeavor. Management has been lulled into complacency by such myths as (1) the injured person is the one who messed up, and (2) everybody is responsible for his or her own safety and the safety of others. Thus, management is totally innocent, or so it thinks.

It is management's prerogative to set up supervisory controls, systems, and standards and to take the lead to ensure that the root causes do not exist.

FAULT FINDING

Failing to identify and rectify root causes before they result in a loss is relying on luck to prevent the next accident. Root causes must be eliminated by adequate controls. People are human beings and have made mistakes in the past and will continue to make mistakes because of the very nature of humanity. Setting the controls will help create an environment in which people are less inclined to make mistakes (mess up).

Example 5

Figure 3.3 is an extract from an accident loss causation analysis, showing the root causes of a breakdown in the system that led to loss. In this example, some of the root causes were identified as poor work planning, lack of motivation, stress, inadequate training, inadequate inspections, no enforcement of procedure, and a failure to analyze risks. Some of the job factors identified were poor maintenance, use of malfunctioning equipment, inadequate corrective action. Inadequate training, and poor inspections and corrective actions were also identified.

These root causes reflected weaknesses in the SMS. There was non-compliance with the hazard warning system, the operator training was generalized and not for specific items of equipment, and no risk assessments of the situation had been done.

WEAK OR NON-COMPLIANCE WITH SMS	ROOT CAUSES
Inadequate Control, Weak SMS	**Personal Factors**
• Non-compliance to standard for reporting and correction of equipment in bad order. • Inadequate training program standards that generalize training for different equipment of the same type. • Non-compliance to standard for reporting damage and operation of barrier. • No risk-assessment standards.	• Driver was exposed to undue demands resulting in stress. (Supervisor's expectations) • Improper motivation to save time and effort. • Lack of skill: • Inadequate initial instruction. • Inadequate practice. (#5 Motor) • Infrequent performance of task. (Tow Motor) • Lack of coaching. (Tow Motor)
	Job Factors:
	• Management does not plan for replacement equipment in the event of mechanical breakdown, resulting in electricians' desire to keep damaged equipment to get job(s) done. • Inadequate employee performance measurement and evaluation. (Proper use of barrier) • Inadequate assessment of maintenance needs and engineering evaluation of changes. (Normal practice to invert cross arm brake adjustment bar) • Inconsistent use of standards, policies, and rules combined with inadequate monitoring of standards compliance. (Reporting damages – electricians and dispatcher) • Inadequate work planning or programming. (Vacation, replacement personnel) • Inadequate assessment of loss exposure – engineering. (2% slope)

FIGURE 3.3 An example of a root cause analysis showing the weaknesses or breakdown in the SMS.

Once these root causes have been identified, the necessary control steps can be instituted to break the sequence of events.

Dan Petersen (1988) summarizes these arguments by saying:

> Perhaps, however, our interpretation of the domino theory has been too narrow. For instance, using the investigating procedures of today, when we identify an act and/or condition that "caused" an accident, how many other causes are we leaving unmentioned? When we removed the high-risk conditions that we identified in our inspection, have we really dealt with *the cause of a potential accident?*
>
> *(p. 17)*

CONCLUSION

Accidental losses are caused by the failure to assess and control the risk. This then leads to the lack of, or a breakdown in, the safety management control system, which then creates root causes, in the form of personal and job factors.

These root causes are the very reason that high-risk conditions exist and that high-risk behaviors are executed, condoned, or tolerated. Root causes are the real problem in the cause and effect of accidental loss. High-risk behaviors and conditions are only the symptoms.

4 Accident Immediate Causes – High-Risk Acts and Conditions

The fourth domino in the CECAL model is the domino representing high-risk behaviors (unsafe acts) and high-risk conditions (unsafe conditions), commonly referred to as the accident immediate causes. Correctly termed, they are the immediate causes of an exposure, impact, or exchange of energy (Figure 4.1).

IMMEDIATE ACCIDENT CAUSES

Bird and Germain (1992) refer to them as the *sub-standard practices* and *sub-standard conditions*. They further explain the immediate causes of accidents:

> The immediate causes of accidents are circumstances that immediately preceded the exposure, impact or contact with a source of energy. They usually can be seen or sensed. Frequently they are called "unsafe acts" (behaviors which could permit the occurrence of an accident) and "unsafe conditions" (circumstances which could permit the occurrence of an accident). Modern managers tend to think a bit broader, and more professionally, in terms of *sub-standard practices* and *sub-standard conditions* (deviations from an excepted standard or practice).
>
> *(p. 26)*

FIGURE 4.1 The immediate causes of accidents, high-risk behaviors, and conditions.

DOI: 10.1201/9781003385943-5

DEFINITION

A high-risk behavior is "the behavior or activity of a person that deviates from normal accepted safe procedure." A high-risk condition is "a hazard or the high-risk mechanical or physical environment." Some major high-risk behaviors and high-risk conditions are as follows.

HIGH-RISK BEHAVIOR

- operating equipment without authority
- failure to warn
- failure to secure
- operating at improper speed
- making safety devices inoperable
- removing safety devices
- using defective equipment
- using equipment improperly
- failing to use personal protective equipment properly
- improper loading
- improper placement
- improper lifting
- improper positioning for task
- servicing equipment in operation
- horseplay
- under influence of alcohol and/or drugs

HIGH-RISK CONDITIONS

- inadequate guards or barriers
- inadequate or improper protective equipment
- defective tools, equipment, or materials
- congestion or restricted actions
- inadequate warning systems
- fire and explosion hazards
- poor housekeeping; disorderly workplace
- hazardous environmental conditions
- noise exposures
- radiation exposures
- high or low temperature exposures
- inadequate or excessive illumination
- inadequate ventilation

Example

In investigating a near-miss incident (undesired event with high potential for loss under different circumstances), the investigator identified three high-risk conditions and three high-risk behaviors that led to the situation that could have caused serious injury.

The high-risk conditions were:

1. The ground had shifted due to natural factors (underground mining).
2. The ground had not been supported for 11 hours.
3. Outdated procedures.

The high-risk behavior that contributed to this near-miss incident were:

1. Timber supports had not been placed in position.
2. Written guidelines issued nine months previously had not been followed.
3. The three supervisors had all failed to warn the employee.

HIGH-RISK BEHAVIOR – FLAWED RESEARCH

The high-risk behavior and high-risk conditions are normally the most obvious events preceding the exposure, impact, or exchange of energy, and consequently, investigations tend to focus on them as being the true cause of accidental loss. The fact that high-risk behavior of an employee is involved also attracts much attention, and because of extensively misquoted statistics, high-risk behaviors are the focus of accident prevention programs.

Jim Howe, CSP, (1998) says:

> The claim that 90% (or similar fraction) of injuries are due to high-risk acts is a repetition of Heinrich's "research." Heinrich's conclusion was based on poorly investigated supervisory accident reports, which then, as now, blamed injuries on workers. He concluded that 88 percent of all industrial accidents were primarily caused by high-risk acts. Companies that sell behavior-based safety programs continue to mislead clients by perpetuating this folklore.
>
> *(p.20)*

Bird and Germain (1992) support this statement by saying:

> Also, an increasing number of safety leaders confirmed the results from research in quality control that 80 percent of the mistakes (sub-standard/high-risk acts), that people make are the result of factors over which only management has control. This significant finding gives a completely new direction of control to the long-held concept that 85–96 percent of accidents result from the high-risk acts or faults of people. This new direction of thinking encourages the progressive manager to think in terms of how the management system influences human behavior rather than just on the high-risk acts of people.
>
> *(p. 26)*

As Jim Howe, CSP, (1998) says:

> This conclusion that 88% of accidents are primarily caused by unsafe acts has been misleading and has led management and others to believe that unsafe acts were the main item to concentrate on to eliminate accidents. In studying the research method that Heinrich, the National Safety Council, and others used to derive these statistics,

it is of vital importance to note that the principal of multiple causes was totally disregarded when these figures were obtained. Heinrich (1959) said, "But *in no case* were *both* personal and mechanical causes charged."

(p. 21)

Cause of Major Importance

What the researchers did, was to consider only the cause of *major importance*. This swayed the statistics. Who was the judge deciding which cause, either immediate or root, was of *major* importance? A description of the methodology used is evidence that the statement that 88% of all accidents are caused by unsafe acts is, in fact, as Howe puts it "folklore."

Heinrich (1959) describes the method used as follows:

> This difference (15%) added to the 73% of causes that are obviously of a man-failure nature, gives a total of 88% of all industrial accidents that are caused primarily by the unsafe acts of persons. Check analyses, made subsequently on a smaller scale, produce approximately the same ratios.
>
> In this research major responsibility for each accident was assigned *either* to the unsafe act of a person or to an unsafe mechanical condition, but *in no case* were *both* personal and mechanical causes charged.
>
> In addition to the research that resulted in the development of the above ratios, other studies have been made, one of chief interest being that conducted by the NSC. This showed unsafe acts for 87% of the cases and mechanical causes for 78%. An analysis made in 1955 of cases reported by the state of Pennsylvania showed an unsafe act for 82.6% and a mechanical cause for approximately 89% *of all accidents.* One reason for the difference in the number of accidents charged to personal or mechanical causes in the three studies described above is that, in the last two, the method permitted both kinds of causes to be assigned for the same accident, whereas in the study first mentioned only the cause of *major importance* was assigned.
>
> Admittedly, judgment must be used in selecting the major cause when a mechanical hazard and an unsafe act both contribute to accident occurrence.

(p. 21)

The "White Paper" survey carried out by *Industrial Safety and Hygiene News (ISHN)* (December 1998) also revealed that 71% of managers surveyed believed that careless employee actions (unsafe acts) caused many accidents (p. 22).

According to *ISHN's* 2015 EHS State of the Nation subscriber survey, much EHS programmatic work in 2015 centered on:

1) Building and/or maintaining a safety culture for organizations (54%).
2) Finding and fixing workplace hazards (48%).
3) Conducting risk assessments and risk prioritization (43%).
4) Tracking safety and health performance measures other than counting injuries and illnesses (38%).

The survey indicated a positive move towards proactive health and safety activities rather than focusing on the high-risk behaviors.

MAJOR SAFETY PARADIGM

Unfortunately, the safety pioneers' method of statistical research has created one of the major safety paradigms of our times, and one of the greatest obstacles to the improvement of workplace safety.

According to Fred A. Manuele (1997), the advice given by Heinrich's causation model has been wrongly focused.

> Safety practitioners have prominently used Heinrich's causation model. Other causation models are extensions of it. However, the wrong advice is given when such models and incident analysis systems focused primarily on characteristics of the individual; unsafe acts being the primary cause of incidents; and measurements devised to correct "man failure," mainly to affect an individual's behavior.

Heinrich also wrote,

> A total of 88% of all industrial accidents … are caused primarily by the unsafe acts of persons.
>
> Those who continue to promote the idea that 88 or 90 or 92% of all industrial accidents are caused primarily by the unsafe acts of persons do the world a disservice. Investigations that properly delved into causation factors prove this premise invalid.
>
> *(p. 30)*

Heinrich et al. (1969) give Dr. Zabetakis's theory of the unplanned transfer or release of energy that causes personal injury and property damage. They quote his explanation as follows:

> Most accidents are actually caused by the unplanned or unwanted release of excessive amounts of energy (mechanical, electrical, chemical, thermal, ionizing radiation) or of hazardous materials (such as carbon monoxide, carbon dioxide, hydrogen sulfide, methane, and water). However, with few exceptions, these releases are in turn caused by unsafe acts and unsafe conditions. That is, an unsafe act or an unsafe condition may trigger the release of large amounts of energy, which in turn cause the accident.
>
> *(p. 32)*

A COMPLEX SITUATION

The term, high-risk behavior, or unsafe act describes a complex situation. High-risk behavior is not merely a worker blatantly defying safety rules and regulations. It is the end action that results from an accumulation of a number of breakdowns and weaknesses in a management system, which was designed to keep the worker safe at work.

Accident investigators should look beyond the high-risk actions uncovered by the investigation and seek the deep-rooted causes for these actions. Many accidents are blamed on the reckless actions or omissions of an individual worker who was directly involved in operational or maintenance work. This is a typical response but is short-sighted and ignores the fundamental failures which led to the accident. These are usually rooted deeper in the organization's design, management, and decision-making functions.

HUMAN FAILURE

Human failure is a broad term often used in accident investigation. There are two main types of human failure, inadvertent failure (error) and deliberate failure (violation).

INADVERTENT FAILURE

Inadvertent failures are classified as mistakes which could be rule-based, or knowledge-based mistakes. Workers do make mistakes, but they are not the prime accident cause. According to Bill Hoyle (2005) in his paper, *Fixing the Workplace, Not the Worker*:

> When you read a newspaper account of an industrial accident it will almost always conclude that the cause of the accident was worker error. In a society largely based on individualism, the idea that worker mistakes are the primary cause of accidents rings true with most people. There is no denying that workers make mistakes. However, in every industrial accident there are almost always several management safety systems involved which may not be readily apparent.
>
> *(p. 3)*

DELIBERATE FAILURE

Deliberate failures are violations. These violations could be routine or could be because of a certain situation, or exceptional circumstance.

EXCEPTIONAL FAILURE

Exceptional failure is where a person attempts to solve a problem in highly unusual circumstances and takes a calculated risk by breaking the rules.

ACTIVE AND LATENT FAILURES

High-risk behaviors are sometimes referred to as active failures. These are errors and violations that have an immediate negative result.

According to the Health and Safety Executive (HSE) (UK) (1999) *Managing Human Failures*:

> Active failures have an immediate consequence and are usually made by frontline people such as drivers, control room staff or machine operators. In a situation where there is no room for error these active failures have an immediate impact on health and safety.
>
> Latent failures are made by people whose tasks are removed in time and space from operational activities, for example, designers, decision makers and managers. Latent failures are typically failures in health and safety management systems (design, implementation, or monitoring). Examples of latent failures are:
>
> • Poor design of plant and equipment
> • Ineffective training
> • Inadequate supervision

- Ineffective communications
- Uncertainties in roles and responsibilities

Latent failures provide as great, if not a greater potential danger to health and safety as active failures. Latent failures are usually hidden within an organization until they are triggered by an event likely to have serious consequences.

(p. 11)

ERRORS

An error is an action which fails to produce the expected result. An error may produce an undesired and unwanted outcome. Human error is commonly defined as a failure of planned action to achieve a desired result. There are four main categories of error.

SLIPS OR LAPSES

Slips or lapses are unplanned actions. They are unintended actions which sometimes occur when the wrong step is taken, or due to a lapse, a step of a procedure or process is not done correctly or is left out.

A *slip* happens when a person is carrying out familiar tasks automatically, without thinking, and the person's action is not as planned, such as operating the wrong switch on a control panel.

A *lapse* happens when an action is performed out of sequence, or a step in a sequence is missed.

MISTAKES

Mistakes are made when decisions and actions are taken and later discovered to be incorrect, although the employee thought they were correct at the time. They are failures in a plan of action. Even if the execution of the plan were correct, it would be impossible to achieve the desired outcome.

Rule-based mistakes happen when a person has a set of rules about what to do in certain situations and applies the wrong rule.

Knowledge-based mistakes happen when a person is faced with an unfamiliar situation for which he or she has no rules, uses his or her knowledge, and works from first principles, but comes to a wrong conclusion.

LATENT ERRORS

Latent errors are problems, or traps hidden within systems, which under certain conditions will contribute to an error occurring. They may lie dormant for some time but given a certain set of circumstances they manifest.

VIOLATIONS

Violations are deliberate deviations from safe work standards and procedures. They can be accidental, unintentional, or deliberate. Violations are rule-breaking actions and are deliberate failure to follow the rules. An example is cutting corners to save time or effort, based on the belief that the rules are too restrictive and are not enforced anyway. The different types of violations are as follows.

ROUTINE VIOLATIONS

Routine violations are identified in most violation categories. Routine violations occur when the normal way of doing the work is different from prescribed rules and procedures. Often routine violations are so common among work teams, that they are no longer perceived as violations or high-risk behaviors. This is called "that's the way things are done around here."

UNINTENTIONAL VIOLATIONS

Unintentional violations occur when rules are written which are almost impossible to follow. This could occur when workers do not know or understand the rules that they are expected to follow. An example of an unintentional violation would be the violation of the speed limit posted in the parking lot of a warehouse which reads 5 MPH (8 km/h), which is far too low for the area, and which is almost impossible to maintain.

PROCEDURAL VIOLATIONS

Procedural violations occur when procedures are purposefully deviated from, ignored, or bypassed. This is often summarized as "failure to follow procedures." The reason for this may be that the procedure is incorrect, outdated, or difficult to follow.

EXCEPTIONAL VIOLATIONS

Exceptional violations occur when an isolated departure from procedure occurs. This type of violation is neither typical of the employee, nor condoned by supervision. Exceptional violations occur in unusual circumstances. In some crisis situations these violations may even be inevitable, especially when it is believed that the violation is necessary to cope with the exceptional circumstances.

SITUATIONAL VIOLATIONS

Situational violations occur when circumstances in the workplace, such as time, pressure, or a sense of urgency, require or encourage employees to violate safety rules. One of the root causes of many accidents has been found to be stress due to having to meet deadlines and production quotas.

THE TOP 10 OSHA VIOLATIONS

During 2021 OSHA issued ~21,000 citations combined in the following categories which were the top ten violations in the US:

1. Fall protection
2. Respiratory protection

3. Ladders, construction
4. Hazard communication
5. Scaffolding, construction
6. Fall protection training, construction
7. Control of hazardous energy (lockout/tagout)
8. Eye and face protection, construction
9. Powered industrial trucks
10. Machinery and machine guarding (OSHA Website, 2022).

OTHER TYPES OF ERRORS

PERCEPTUAL ERRORS

Perceptual errors occur when an operator's sensory input is degraded, and a decision is made based upon this faulty information. Unclear signals or mixed messages can cause perceptual errors to occur. Conflicting demands, an accident root cause, is an example of a perceptual error.

RULE-BASED ERRORS

Rule-based errors are when a worker applies written or memorized rules to deal with an unfamiliar situation. Rule-based errors are situations where the use, or disregard of a particular rule, or set of rules, results in an undesirable outcome. Some rules that are appropriate for use in one situation may be inappropriate in another. The misinterpretations of rules, or deviations from prescribed procedures, lead to mistakes. This may happen when changes in the situation prevent an individual from relying on skills.

SKILLS-BASED ERRORS

Skills-based performance errors are situations in which workers perform a task with little conscious thought. This is usually the result of extensive experience with a given operation. These are actions people do, almost without thinking, like riding a bicycle or typing a letter.

When operating in a skills-based performance mode, most mistakes are due to inattention. These errors occur in experienced situations. They occur in the worker's execution of a routine, often well practiced task. They could be due to a memory lapse or slip of action. In a skills-based mode, workers rely on work experience of having done the same task with little or no attention over the years.

DECISION-BASED ERRORS

Choosing the wrong course of action may result in an unsafe situation. Mistakes are decision-making failures. They arise when workers do the wrong thing, believing it to be right.

KNOWLEDGE-BASED ERRORS

Knowledge-based performance relies on a worker's understanding of a task. Many errors result from flaws and weaknesses in that understanding. This is also known as knowledge-based mode, or lack of knowledge-based mode. Because knowledge-based performance relies on an individual's knowledge-based performance, when workers do not know what they are doing, such as when faced with totally unfamiliar situations, they rely on existing knowledge to help them. They look for patterns and apply a remedy they have learned from other tasks to the situation facing them. Sometimes a wrong step is applied because of this lack of knowledge.

ERROR CHAIN

There is a concept in aviation called an error chain. An error chain can be defined as: "a sequence of minor mistakes leading to a disaster, or a series or chain of events culminating in a loss." This means it is not a single error, not two errors, but several successive errors in judgment or execution. If any of them had been avoided, there would have been no loss.

The error chain refers to the concept that many contributing factors typically lead to an accidental loss, rather than one single event. Each link in the error chain is an event that contributes to the loss. This supports the principle of multiple causes.

The error chain can be just a single link where just one mistake can end in disaster, or it can be many links where things all have to line up perfectly for the loss to happen. Breaking the error chain is when someone intervenes to stop a chain of events that, if allowed to continue, would ultimately result in an unplanned loss.

VITAL FACTOR

It is obvious that the high-risk environmental conditions and high-risk behavior of people (unsafe acts, high-risk behavior, or at-risk behavior as they are sometimes called) are always present in the causation of accidental loss. As the CECAL model shows, they are a factor in the sequence of events that lead to accidental loss. If they were removed, the exposure, impact, or contact and subsequent loss would not occur. Early safety philosophers relied entirely on attempting to remove the high-risk behavior and high-risk condition and, due to finger-pointing exercises, tended to (and still do) focus on the high-risk behavior of the individual without really delving into the reason (root cause) for that person's high-risk behavior.

Heinrich et al. (1969) calls the immediate causes the symptoms:

> These are the factors in the accident/incident sequence that have historically been called the most important ones to attack. They are also the factors that receive the bulk of attention in governmental safety and health inspections around the world.
>
> *(p. 26)*

As quite rightly stated, most safety legislation focuses on safe physical conditions and safe behaviors of employees rather than focusing on management systems, controls, and checks and balances that are integrated into the day-to-day management to ensure that the conditions are maintained to an accepted safe level and that behaviors are controlled.

SAFETY PARADIGM

Despite current safety philosophies, we still appear to believe that the majority of accidental losses are caused by the high-risk behaviors of people. This idea was quoted extensively in Heinrich's book *Industrial Accident Prevention*, which was first published in 1929. Even in the fifth edition, which was co-authored by Dan Petersen, PE, CSP, and Nester Ross, D.B.A., the ten original axioms on industrial safety are stated. Axiom 2 reads, "The unsafe acts of persons are responsible for a majority of accidents" (p. 21).

THE 88%, 10%, 2% RATIO

The first edition of Heinrich's book stated that 88% of all accidents were caused by unsafe acts, 10% were caused by unsafe conditions and 2% were attributed to acts of providence. The National Occupational Safety Association (NOSA) (in their *Questions and Answers on Occupational Safety and Health* series of books) further promoted these statistics. In the *Advanced Questions and Answers Manual* (1988) question 9 asks:

> What is the key to accident prevention?
>   Answer: If we were able to eliminate all unsafe acts and unsafe conditions, about 98% of all accidents would be prevented. Unsafe acts cause approximately 88% of all occupational accidents. Unsafe conditions cause approximately 10% of all occupational accidents. Acts of providence or natural phenomena cause approximately 2% of all occupational accidents.
>
> *(p. 6).*

In subsequent publications such as the *Health and Safety Training, General Course Manual* (1994) the following statement appeared, "Human error/inefficiency – 88% i.e., the human factor; Unsafe conditions – 10% i.e., the engineering factor; Acts of nature – 2% i.e., the inevitable" (p. 23).

High-risk behavior and *conditions* are further defined in paragraph 14.3 as "acts, which are willful or negligent or knowingly ignoring set standards will sooner or later result in an accident or incident" (p. 25).

In paragraph 10, *unsafe conditions* are defined as "any variation from accepted safety standards, which may be the cause of incidents and/or accidents" (p. 21).

Du Pont seminars teach that 96% of all injuries are caused by unsafe acts, "Because unsafe acts cause 96% of all injuries ..." and justify this statement by declaring: "When we conducted a 10-year study of all serious injuries occurring at Du Pont sites – at offices, on refineries, in transportation, at all kinds of plant sites – we learned that 96% of our injuries were caused by the unsafe acts of people and poor work practices."

Focusing on the high-risk behavior alone will not necessarily mean the interruption of the loss causation sequence. Dr. Mark A. Friend (1997) asks the question:

> Since 90% of all accidents and adverse incidents are due to human error, should safety efforts be focused on correcting these errors? The answer: False. Although 90% of all accidents and adverse incidents may, in fact, be caused by human error, safety efforts must also encompass engineering principles. When asked whether the engineering or human relation's school should be emphasized, Fred Manuele says, "A curse on both their houses." A seminar leader once suggested that since 95% of all safety problems are due to human error, 100% of solutions should address the human element. His reasoning: by taking such action, nearly all problems will be covered. Such thinking is naive and dangerous. Addressing the human element can solve some problems. But ignoring engineering solutions is akin to fixing the problem with a short-lived solution that will fail when the slightest change occurs.
>
> *(p. 36)*

WHY?

Safety practitioners, management, and union management as well as fellow employees have asked this question a thousand times, "Why do workers commit unsafe acts?" There is no simple answer to this question and each accident investigation that has involved high-risk behavior has always indicated that there were reasons and motivation that caused the person to commit the high-risk behavior.

Dan Petersen (1998) attempts to give an explanation by showing how workers personally benefit from working unsafely. He gives four reasons:

1. The advantages and satisfaction to be gained by the worker at the particular moment seem greater to him than the disadvantages and dissatisfactions.
2. The unsafe act "makes real sense" to the employee. If he is challenged, he will explain to the supervisor exactly why he thinks his way is the most sensible way to do the job. Typically, the older employee will justify himself by saying that he has been doing it that way for years.
3. The unsafe act actually gives the worker personal satisfaction; it may attract the attention of co-workers; gain their approval and admiration; give him either the thrill of taking a chance or the satisfaction of bucking authority – or even paying back an imagined grudge; it may make him feel daring; and it may involve many other personal incentives.
4. To the worker, his "unsafe" act may be perceived as having definite job-related advantages – advantages that include either such monetary incentives as getting his job done sooner, thus increasing his work-output and his take-home pay – especially if he is on piecework pay – or personal incentives such as avoiding extra effort or fatigue and having more "personal control" over product quality (p. 232).

Having worked with employees in industrial situations for a number of years, my personal conclusions are that people commit high-risk behavior because they are (we are) human beings. When questioning motorists as to why they travel at 60 mph (110 kmh) instead of the limit of 55 mph (90 kmh), they all generally agree that the advantages of the risk generally outweigh the disadvantages. They also rationalize that breaking the rules "a little bit" will never hurt anyone. I was once told that an African philosophy is that if you are milking the cow and take a sip of the milk, the cow will not die.

COMPLACENCY AND CONDONED PRACTICE

Numerous accident investigations found that the high-risk behaviors were not unique to the accident and had been executed many times previously. High-risk conditions do not merely pop up and cause an accident but have also been there for a long while. Complacency and condoned practices are what cause high-risk behavior and high-risk conditions. In numerous accidents, fellow workers, supervisors, and others ignored the high-risk behavior. Once the high-risk behavior resulted in an exposure, impact, or contact and an accidental loss such as injury, the employee was immediately confronted with his or her high-risk behavior. The same happens with high-risk conditions.

DISCIPLINE

Another safety paradigm is disciplining employees for committing high-risk behaviors. I asked the question, "How can an employee be disciplined for committing a high-risk behavior when one high-risk condition is present in the workplace?" Safety management is a two-edged sword and both sides should be viewed equally. Allowing high-risk behaviors to go unheeded is tantamount to committing a high-risk behavior, and one deviation from accepted workplace health and safety standards is also unacceptable. Besides, will disciplining that employee fix the safety problem?

Example

In the 3C accident investigation there were found to be numerous high-risk behaviors and conditions. There was a failure to warn, operating at unsafe speed, safety devices made inoperable, defective equipment, improper positioning, people not following procedures, defective equipment, absence of a warning system, and a hazardous environment. Figure 4.2 shows the high-risk behavior and conditions listed under the fourth domino.

Multiple Causes

The principle of *multiple causes* states that *accidents and other problems are seldom, if ever, the result of a single cause.* This means that there is always more than one cause for an accident, and seldom will an accident be because of a single isolated high-risk behavior or high-risk condition. Normally, there are both behaviors and conditions. If one counts only the high-risk behaviors, one could easily derive the statistic that the majority of accidents are caused by the high-risk behaviors.

High-risk Behavior	High-risk Conditions
The controller was not checked.	The transformer was running too hot.
The machine was operated at an unsafe speed.	Inadequate ventilation.
Safety devices had been rendered inoperable.	The stop limit switch was not in place.
The operator was not aware of the danger of the faulty sensor	The environment was hazardous and there was inadequate lighting
The computer indicated the wrong position.	No warning system existed.

FIGURE 4.2   The 3C accident investigation.

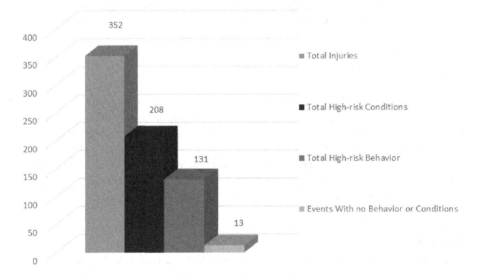

FIGURE 4.3   Immediate cause analysis (year 1).

Research (Company XYZ)

Over a three-year period an analysis was made of injury reports as to what percentage of injury-causing accidents were caused by high-risk behavior and what percentage by high-risk conditions. The principal cause of the accident was also used to derive these statistics.

Findings

During the first year a total of 352 injuries were recorded, with 208 being the result of high-risk behaviors and 131 as a result of high-risk conditions. This meant that some 40% of the accidents were caused by the high-risk conditions and 60% by the high-risk behaviors. Figure 4.3 shows the analysis produced.

In year 2, the picture changed with 388 total injuries reported. One hundred and thirty were the result of high-risk behavior and 180 the result of high-risk conditions. Rounding off the figures, 34% of accidents were caused by the high-risk behavior and 47% by the high-risk conditions. In year 2, 78 injuries had neither behaviors or conditions listed and were therefore removed from the equation (Figures 4.4 and 4.5).

During the third year, the same study was undertaken, and the findings were as follows: There were 341 injuries recorded because of accidents, 114 were because of the high-risk behavior and 172 because of the high-risk conditions. These then represented 34% caused by high-risk behavior and 50% by high-risk conditions. Figure 4.6 shows this research.

GENERIC STATISTICS

Unless a company has conducted research as to what are the immediate causes of their loss-producing accidents, it would be unwise to quote generic statistics, which are slanted toward either the high-risk behavior or the high-risk condition. It is also opportune to acknowledge that high-risk behavior and high-risk conditions are factors in the cause and effect of accidental loss.

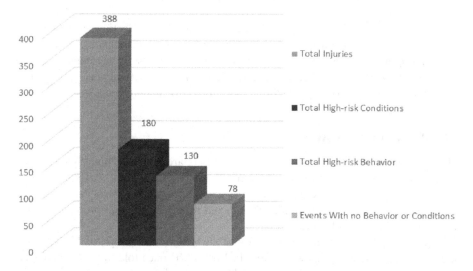

FIGURE 4.4   Immediate cause analysis (year 2).

YEAR 2				
Total Injuries	Total High-risk Behavior	Percentage of High-risk Behavior	Total High-risk Conditions	Percentage of Conditions
388	130	33.51%	180	46.40%

FIGURE 4.5   Immediate cause analysis percentages (year 2).

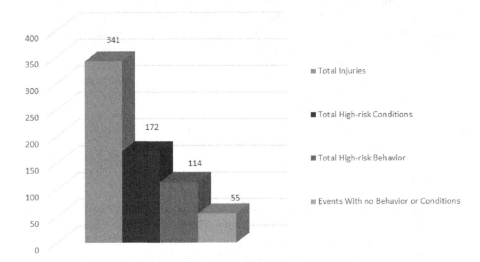

FIGURE 4.6   Immediate cause analysis (year 3).

The percentages should be compiled by accurate data collection and statistical analysis and be pertinent to a particular workplace. Generalizing these statistics and quoting high-risk behavior as the ultimate and only cause of accidents wrongly points the finger at the worker who is only a pawn in the safety management game. The injured employee is only a victim of a failure of the management system.

ACCIDENT ROOT CAUSES

High-risk behaviors and high-risk conditions exist or are caused by underlying causes in the form of personal and job factors (root accident causes). Rather than treat the symptoms of a failure of the management system, immediate accident causes should be used as a means of probing, identifying, and rectifying the real problems in the form of personal and job factors.

The work environment impacts heavily on the behaviors of people and can dictate how they act. In discussions with Frank E. Bird, Jr. he once told me, "If a manager cleans up the work environment, he also cleans up the thought processes of the people in that environment." There are reasons, root causes, for high-risk behavior and there are also reasons for high-risk conditions. The best and only way to eliminate the immediate causes is to find and eliminate the root causes.

EXAMPLE

In Case Study 4 (*Questions and Answers on Effective Accident/Incident Investigation* 1990), the injured employee's recollections of his actions, a combination of his high-risk behavior which created a high-risk condition, which led to devastating personal consequences for him, are related as follows:

Then came the most terrible day in my life, 21st September, at 2.30 pm. I was working on the steering clutches of a piece of heavy earth- moving equipment. To replace core lead inserts on a steel band I had to grind off some old welding. I was working on a rough and ready construction site and the only grinder (angle grinder) had been bolted onto a box and the box had been filled with rocks for added "safety" and stability as a base.

I could not reach inside the break bands to grind with the guard in position. Without thinking, I removed the guard. I did not realize that I was adding more danger to an already hazardous situation. Suddenly, as I commenced working, the wheel disintegrated. I felt a terrible pain in my stomach. A wedge-shaped piece of the wheel sliced its way through my abdominal muscles and into my abdominal cavity. It caused a 4-inch (12 cm) long wound in my abdominal muscles. It then left my body through a 3-inch (8 cm) gash in my back and smashed into my lower leg.

(p. 51)

This accident left the employee on the construction site for some two hours and he only reached the hospital two hours later. Four hours and ten minutes after that, the surgeon arrived, and he was in the operating theater for 3 hours and 30 minutes. The employee's foot bones were smashed, and his foot had to be amputated. He then spent a further 28 days in hospital and has had tremendous trouble with his artificial limb and doctors even suggested a second amputation at the knee.

BEHAVIORAL-BASED SAFETY

Dr. Thomas R. Krause (1997) defines behavioral-based safety as:

The phrase *behavioral based safety* refers strictly to use of applied behavior analysis methods to achieve continuous improvement in safety performance. These methods include identifying operationally defining critical safety-related behaviors, observing to gather data on the frequency of those behaviors, providing feedback, and using the gathered data for continuous improvement.

(p. 30)

Dr. Krause continues by saying:

Wherever it appears in this book, the statement that most accidents are "caused by" at-risk behavior, the type of cause referred to is known as the final common pathway. Those critical behaviors are "produced" by management systems of the site and therefore decreasing them is the key to accident prevention.

(p. 30)

In discussing behavioral analysis, he continues by saying:

Behavioral analysis is one of the basic tools of the employee-driven safety process. Most safety problems involve someone's behavior. Behavior is controlled by antecedents (things that trigger behavior), and by consequences (the things that both follow from behavior and influencing it).

(p. 30)

Using a procedure based on safety observations, Krause quotes a case study of outcomes at a paper mill employing 425 employees. The case study is summarized by his quotation:

The mill realized a 47% reduction in injuries in the first 2 years of their process (p. 50).

He concluded that by focusing on behaviors critical to safety performance, this paper mill achieved a steady decrease in reportable injuries. He further stated that:

Ongoing peer to peer observations and feedback about safety performance has become part of the mill's culture: it's what the employees talk about.

(p. 50)

Dr. Krause also motivates behavior-based safety by saying:

The reason to focus on behavior is that when an incident occurs, behavior is the crucial final common pathway that brings other factors together in an adverse outcome. Therefore, ongoing, upstream measurement of the sheer mass of these critical at-risk behaviors provides the most significant indicator of workplace safety.

(p. 19)

He describes the behavior-based safety improvement process as:

- identify critical behaviors
- identify root causes
- generate potential actions
- evaluate possible actions
- develop action plans
- implement action items (pp. 79–89)

His safety improvement process model indicates that if the problem is not solved, the identification of critical behaviors must be reverted to and if the action items are complete, reverts to identifying root causes and generating the potential actions. If the actions are incomplete, re-implement the action items.

The success rates in the paper mill case study discussed by Krause measure the recordable injury rate decrease and the number of employees injured. If measuring the degree of injury is accepted as a good indication of safety, then the results of behavioral-based safety according to Dr. Krause are significant. It has often been quoted, however, that the measurement of injury, and specifically a certain degree of injury, is not reliable, as the injury is largely dependent on the luck factors 1, 2, and 3 as will be discussed in further chapters of this publication.

Safety Success

James Tye, director-general of the British Safety Council, was quoted as saying that (injury) frequency rates were measurements of failure, rather than success. He further explained that since the underreporting of accidents and injuries could be as high as 80% in some U.S. industries, he concluded that frequency rates were irrelevant in measuring safety performance.

According to Bird and Germain, measurement of management control is a far better and reliable measurement of safety performance. Bird and Germain also quote Charles E. Gilmore, who, during an address at the National Safety Congress, said:

> What is the sense in measuring if the loss must occur before you can act? That's reaction, not control.
>
> *(p. 49)*

Is Dr. Krause measuring the success of the behavior-based safety systems merely by measuring the degree of injury rather than degree of control?

AT-RISK BEHAVIOR

Dr. E. Scott Geller (1996) also quotes Heinrich:

> Heinrich's well-known law of safety implicates at-risk behavior as the root cause of most near-hits and injuries. Over the past 20 years, various behavior-based research studies have verified this aspect of Heinrich's Law by systematically evaluating the impact of interventions designed to lower employees' at-risk behavior. At-risk behaviors are presumed to be a major cause of a series of progressive, more serious incidents.
>
> *(p. 76)*

Geller refers to "near-hits," but the American Society of Safety Professionals' technical dictionary does not define the concept *near-hit* but does define the concept *near-miss*.

Based on this approach of presuming high-risk acts to be the major cause of serious incidents (accidents) as stated by Geller, Jim Howe, CSP (1998), says that:

> The behavioral safety approach is biased. It ignores hazards and risks and focuses on "critical worker behaviors," which would permit working in a hazardous environment. This almost always leads to the implementation of low-level controls, safety procedures and personal protective equipment instead of more effective engineering control.
>
> *(p. 20)*

Geller (1996) also warns that safe acts should not be ignored, and that corrective feedback should be used.

> Since adverse behaviors contribute to most if not all injuries, a Total Safety Culture requires a decrease in at-risk behaviors. Organizations have attempted to do this by targeting at-risk acts, exclusive of safe acts, and using corrective feedback, reprimands, or disciplinary action to motivate behavior change.
>
> *(p. 86)*

Differences

Jim Howe, CSP (1998), defines the difference between a systematic approach and the behavioral systems approach as follows:

> There is a substantial difference between a systematic approach to workplace health and safety and a behavioral safety approach. The system approach takes an objective and unbiased view of the workplace by:
>
> 1. Identifying hazards.
> 2. Estimating the level of risk for each hazard.
> 3. Controlling hazards according to the hierarchy.
>
> Behavioral-based safety programs appeal to many companies because they make health and safety seem simple, do not require management change, focus on workers, and seem cheaper than correcting health and safety hazards.
>
> *(p. 20)*

Behavior and Conditions

High-risk behavior and high-risk conditions are factors in the accident sequence. Focusing on only one factor in a multiple causation sequence would be ineffective and therefore a holistic control approach is required.

The high-risk behavior and high-risk conditions will continue and will always be the event that result in an exposure, impact, or exchange of energy, which in turn causes the loss.

The high-risk behavior and high-risk conditions will remain the obvious causes of the accidental loss but are clearly only the symptoms of other greater failings within the organizational system. To have a balanced approach, employees, supervision, and management have specific duties in safety management. They are as follows.

Management

Management is ultimately responsible for health and safety and their responsibility is to:

- Set objectives and develop a policy for health and safety.
- Bring operations in line to comply with applicable legislation.
- Delegate responsibility and authority to those at various levels in the organization for certain duties and functions within the SMS.
- Ensure that health and safety information is an integral part of training and operations.
- Ensure that contractors comply fully with company and other applicable safety regulations.
- Maintain an industrial hygiene monitoring system.
- Set a good example by attending health and safety meetings and taking action on accident investigation reports.

Various other levels of supervision also have specific roles to play in safety management. The employees' roles are as follows:

- Must work according to health and safety standards and procedures.
- Must observe health and safety rules and regulations.
- Must report hazardous conditions and high-risk practices.
- Develop and practice good habits of hygiene and housekeeping.
- Must use personal protective equipment properly and maintain it in good order.
- Report all injuries and near-miss incidents.
- Assist in developing safe work procedures.
- Make suggestions for improving health and safety conditions and procedures.

CONCLUSION

The failure to identify hazards and assess and control workplace risks creates a breakdown in the management safety control function, which creates the root causes of personal and job factors, which in turn lead to high-risk behavior and high-risk conditions.

SYMPTOMS AND NOT CAUSES

High-risk behavior and high-risk conditions are the symptoms of the loss-causing situation and should be used to identify and eradicate the root, underlying, causes. High-risk behavior or a high-risk condition will inevitably lead to an exposure, impact, or contact with a source of energy, which will lead to loss.

5 Luck Factor 1

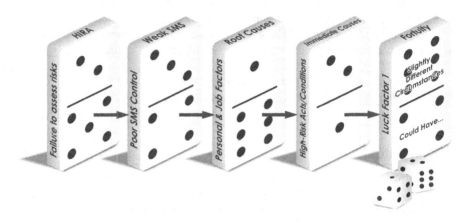

FIGURE 5.1   Once a high-risk situation exists, Luck Factor 1 determines the outcome.

FORTUITY

As one seasoned safety professional put it, "In forty years of working in safety, no one has yet been able to explain to me what determines the difference between a near-miss incident and an accident that injures workers."

THE SEQUENCE SO FAR

Failure to assess the hazards and control the risk triggers off poor controls in the form of an inadequate health and safety management system (SMS). This causes personal and job factors. These root accident causes create a climate that breeds high-risk behaviors and high-risk conditions. The stage has been set, and invariably the next event, because of the high-risk condition or action, is either an exposure, impact, or contact with the source of energy or a near-miss incident, depending on Luck Factor 1.

It would be appropriate at this stage to redefine the terms *accident* and *near-miss incident*.

ACCIDENT

An accident is, "an undesired event, which causes harm to people, damage to property or loss to process," according to Bird and Germain (1992) (p. 18).

NEAR-MISS INCIDENT (NEAR-MISS)

Bird and Germain (1992) define a near-miss as, "an undesired event, which, under slightly different circumstances could have resulted in harm to people, damage to property or loss to process" (p. 19).

                                                     DOI: 10.1201/9781003385943-6

No Logical Explanation

It is therefore concluded that there is no logical explanation why some high-risk behaviors end up as accidents (with loss) and the same high-risk behavior, under slightly different circumstances, ends up as a near-miss incident (with no loss). The difference is often determined by good fortune or luck. The outcome is fortuitous.

LUCK FACTOR 1

In referring to the CECAL model, it is proposed that Luck Factor 1 (fortuity) determines the outcome of a high-risk behavior, high-risk condition, or combination thereof. The outcome can either be an exposure, impact, or contact with a source of energy or merely a near-miss incident. A near-miss incident is an undesired event, which, under slightly different circumstances, could have caused a loss. It has potential to cause a loss. What determines the *slightly different circumstances,* the difference between a loss and no loss?

Shift in Time or Position

Some define the Luck Factor as being in the wrong place at the right time. Some explain it as a result of timing, positioning, circumstance, unexplained differences, and other factors which were not planned, nor could be explained.

OSHA defines a near-miss as, "an incident in which no property was damaged, and no personal injury was sustained, but where, given a slight shift in time or position, damage or injury easily could have occurred."

Others explain a near-miss incident as an unintentional incident that could have caused damage, injury, or death but was narrowly avoided. Yet they do not explain why the event was narrowly avoided.

Another way to put it is that, on the other hand, the absence of an accident does not mean there was no violation – it may only reflect the employer's good fortune.

Positioning

At an aluminum smelter casting wheel, the team of workers usually stood at the north side of the wheel during the pouring of the molten metal. For some unknown reason they took up position on the south side for the second pour of the day. As the crane lifted the heavy crucible of molten metal, the crane cable slipped, and the crucible dropped down onto the casting table. The jolt sent molten metal flying all over the north side of the casting bay where the team normally assembled. Despite almost a half a ton of metal being sprayed all over, no one was injured. No one could explain why, at that moment, the team had decided to gather at the south side of the wheel.

MEASUREMENT

Dan Petersen (1998) talks on safety measuring systems as follows:

> In the early days of safety, accident measures, such as the number of accidents (injuries), frequency rates, severity rates, and dollar costs, were used to measure progress.

Although it became clear long ago that these measures offer little help, they continue to be used today. Why should we consider other measures? "Results" measures nearly always measure only luck – unless they concern a huge corporation that generates thousands of injuries.

(p. 37)

ENERGY SOURCE

What differentiates an accident from a near-miss incident is an *exposure, impact,* or *contact* with a source of energy or substance greater than the threshold limit of the body. This means that there was an exchange of energy that caused the loss. The contact with this source of energy is always as a result of a high-risk behavior or high-risk condition. In thousands of documented cases, high-risk behaviors have been committed and there has been no contact or no consequence whatsoever.

UNDER SLIGHTLY DIFFERENT CIRCUMSTANCES

Similarly, the same high-risk behaviors have been committed, and under slightly different circumstances, have resulted in contact with a source of energy and a subsequent loss to either people, property, or the process. The only plausible explanation is that Luck Factor 1 determines the outcome of high-risk behaviors and conditions.

NO LUCK IN SAFETY

This luck factor theory has often been challenged with the statement "There's no such thing as luck in safety!" This is agreed, there is no luck in safety, but, the outcomes of an undesired event cannot be accurately predicted nor controlled and are therefore fortuitous. This has been stated by safety pioneers over and over, yet organizations are convinced that because there was no injury there was no accident. No blood, no foul.

VAAL REEFS DISASTER

My first safety training in the USA was conducted at a large mine. While I was introducing myself to the group, a young miner in the audience stood up and challenged me, "Hey you, Crocodile Dundee, what the heck can you teach us about safety?" Being somewhat taken aback by this statement I first inquired why he called me Crocodile Dundee. On hearing my accent, he thought I was from Australia and not South Africa. I then asked him what he meant by his statement. He continued, "You 'fellas' in South Africa have just killed 105 people at a mine in South Africa where a 12-ton (11,000 kg) locomotive fell down the mine shaft and collided with a cage full of miners. It took the cage down to the bottom crushing it completely and killing all 105 occupants." There was a terrible silence in the room, and he continued, "If those types of accidents happen in South Africa, what can YOU teach us about safety?"

THE LEVEL 72 HORROR

He was absolutely correct. On May 10, 1995, at the Vaal Reefs No. 2 main shaft, a 12-ton (11,000 kg) locomotive pulling a man carriage fell down the mine shaft. At the same time a fully loaded man-cage was moving slowly down the shaft. At ~120 ft (40 m) below the 56 level the collision took place and the cage plummeted 1,650 ft (500 m) to the shaft bottom, where it was crushed by the locomotive. The 25 ft (7 m) high, two-deck cage was crushed down to 5 ft (1.5 m). *The Sunday Times* (1995) report entitled "The level 72 horror" tells the story:

> A free falling 12-ton (11,000 kg) locomotive glanced against the lift and pieces of metal smashed through its top, hitting Mr. Quluba on the head. His neck was fractured, and one metal piece caused a fast bleeding wound just above his forehead. Luckily, he survived, and the lift made its way to the top.
>
> But 105 other miners were not so lucky. They died when the locomotive and its carriage smashed into their cage on level 56, sending it plummeting to the bottom at 70 mph (120 km/h).
>
> *(p. 9)*

On the same page, Peter DeIonno continued with the story:

> Evidence will be presented to the disaster inquiry that barriers intended to stop underground trains from approaching the shaft mouth were often removed at the mine to speed up handling of rail carriages. In free fall, the train took 3 seconds to catch up 165 ft (50 m) down with a fully loaded two deck man-cage lift that had just passed downwards at 35 mph (16 m) a second. The National Union of Mineworkers has declared May 17 a day of mourning and is planning marches to protest against poor safety standards at mines.
>
> *(p. 9)*

GUESS WE WERE LUCKY

The young miner certainly had a point and had every reason to question my ability. I then addressed the group, which consisted mainly of underground miners with an average of 17 years' experience. I asked them if a locomotive had ever fallen down one of their mineshafts. There was an uncomfortable silence in the room, which was eventually broken by an old timer who stood up and said, "Yes, in 1960 we dropped an 8-ton (7,250 kg) locomotive down the main shaft." I asked him how many people had died as a result of this accident. He replied, "None, nobody was injured." I asked, "Why?" and was greeted by silence. I then repeated my question. "At Vaal Reefs a loco fell down a shaft and 105 people died. Here, at this mine in 1960, a loco fell down a shaft, the same as at Vaal Reefs, but no one died and nobody was even injured, why?" There was a silence in the room and one of the old-time miners stood up, looked around the room, and said, "Guess we were lucky." How right he was.

They were lucky that the cage was *above* the level from which their locomotive plummeted. The Vaal Reefs scenario was clearly an example of bad luck, the cage having been positioned *below* the level from which the locomotive fell. At the mine in 1960 the cage was *above* the level from which the loco plummeted, so it fell down the shaft without colliding with the cage.

EXAMPLES OF LUCK FACTOR 1

Bearing the Vaal Reefs level 72 disaster in mind, certain research was carried out at an underground mine in an effort to prove that Luck Factor 1 determines the outcome of a high-risk behavior or a high-risk condition. This particular mine had never had a disaster similar to Vaal Reefs, but upon investigation found that it had had three opportunities for a similar disaster.

ACCIDENT 1

In this particular accident, the report stated "The rail came down hitting the end of the transporter, knocking them over the scotching chains and down the shaft..." There were no injuries as a result of this 1,000-lb (450 kg) vehicle falling down the mine shaft.

ACCIDENT 2

A few years later another vehicle fell down the shaft. "As tension was added by the hoist, the cage broke loose and sprang up 6 to 8 feet (2-2.6 m) overturning the battery onto the station and the flat (car) fell down the shaft."

ACCIDENT 3

A similar accident occurred three years later, "This caused the leading flat (flat car) to tip up, come uncoupled and roll into the no. 2 compartment of the shaft. This occurred very quickly and was over in a matter of only a few seconds."

The above three accidents did cause property damage and minor business disruption but there were no injuries and no fatalities primarily due to Luck Factor 1. The circumstances proved to be fortunate because none of the items fell on a cage full of miners. The potential for an accident such as the Vaal Reef's disaster was, however, high.

WARNINGS

S. L. Smith (1994) says that:

> If enough near misses occur, the question is not, "will an actual accident ever happen," but "when will it happen.
>
> *(p. 33)*

HEED THE WARNINGS

Many call the near-miss incidents "warnings" and quite rightly so. Safety management is perhaps the only management science where numerous warnings are given before an exposure, impact, or contact and loss occurs. The only reason that an exposure, impact, or contact occurs, or does not occur, is because of the luck factor. The warnings, in the form of near-miss incidents, should be heeded, nevertheless.

ACCIDENT INVESTIGATION

In investigating an accident in which a worker climbed a poorly supported electric utility pole and was killed when both he and the pole fell to the ground, it seemed a clear-cut case that the worker had committed a high-risk action. Thorough investigation and some five days of gathering further evidence showed that a fellow worker had been doing exactly the same as the victim except on a different pole. This pole was half a mile (0.8 km) distant from the fatality site. Investigation showed that this pole was also not completely compacted. The worker had ascended the pole on numerous occasions and had completed the running of the conductors through the isolators.

When trying to analyze why the one pole fell, killing its climber, and the other pole did not, one can only conclude that the electrician on the distant pole was lucky.

NEAR-MISS REPORTS

In reviewing monthly near-miss incident reports from employees, the luck factor became very apparent. Near-miss incident 1 was as follows:

> While watching the controlled explosion demolition of an old building, an employee who was behind the safety barrier with other workers, ducked just in time when a chunk of flying concrete from the blast headed straight for him.

This near-miss incident shows that luckily the worker ducked a split second before the chunk of concrete would have hit him.

The next near-miss incident not only indicates the Luck Factor but also the high potential for loss. Near-miss incident no. 2 is as follows:

> We were trying to open the underground man-way shaft doors, to go to the surface. We could not open the door because the door to the shaft and the door to the station were open. We finally got the door open, and the pressure of the air pulled me in toward the shaft. I almost ended up falling down the shaft had it not been for the wire.

Once again, the luck factor determined the difference between a contact and no contact and in this case, circumstances were slightly different, and the man was not sucked into the shaft.

Further Near-miss Incident Examples

- An employee tripped over an extension cord that lay across the floor but avoided a fall by grabbing the corner of a desk.
- A pry bar which was left on the bottom of a mill under repair flew through the air like a missile when a loosened liner fell onto it while the mill was

being rotated. The heavy bar missed workers in the immediate vicinity and neither injury nor damage was sustained.

- An outward-opening door nearly hit a worker who jumped back just in time to avoid a collision with the door.
- Instead of using a ladder, an employee put a box on top of a drum, lost balance and almost stumbled to the ground. Although the employee was shaken, there was no injury.
- An employee tested the brakes at the beginning of the shift, and they checked out ok. As he approached another vehicle, he hit the brakes and they did not work, and he narrowly missed the vehicle.
- While walking from his car to the stairs, an employee was almost struck by a fast-moving pickup.
- A miner was pulling out hoses to set up the jack leg and the hoses hung up making the miner mad. He pulled really hard and lost his balance and fell down. He was not hurt.

ACCIDENT RATIOS

There have been numerous accident ratios proposed over the years. Heinrich (1959) first proposed the concept that there were some 300 narrow escapes to every serious injury, in 1931. The Frank Bird Ratio of 1966 expanded on this theory and showed 600 near-miss incidents with no visible sign of loss for every serious injury. The Tye/ Peterson Ratio estimated 400 undesired events for every serious injury. The main question is what determines the difference in outcomes of these undesired events? What differentiates between contact and non-contact? The only logical answer is the luck factor.

Tarrants (1980) summarizes the luck factor by saying:

> The study of errors and near accidents usually reveals all those situations that result in accidents plus many situations that *could* potentially result in accidents but that have not yet done so.
>
> *(p. 117)*

ACCIDENT RATIO CONCLUSION

There have been accident ratios calculated in the past and there will be new accident ratios calculated in the future. In generalizing all accident ratios, it can be said that for every 1 serious injury there are *some* minor injuries, *more* property damage accidents and *plenty* of near-miss incidents which, under slightly different circumstances (luck, chance, fortune), *could have* resulted in accidents with loss (Figure 5.2).

SUMMARY

Many organizations rely on this luck factor rather than on control for reducing the number and degree of their accidental losses. A near-miss incident reporting and assessing program is vital to an organization as it determines what *could have*

FIGURE 5.2 The updated accident ratio. (From McKinnon, Ron C. 2012. *Safety Management, Near Miss Identification, Recognition, and Investigation.* Model 2.4. Boca Raton, FL: Taylor & Francis. With permission.)

happened under slightly different circumstances. It is important to assess the potential, severity, and frequency of near-miss incidents, as well as accidents that produce a loss.

LUCK OR CONTROL?

In discussions with management who venomously defends its lack of, or few injury accidents, I always ask the question, "Are you in control or relying on luck?"

6 Near-Miss Incidents – Under Slightly Different Circumstances

LOSS OR NO LOSS

So far, the cause and effect of accidental loss (CECAL) sequence has shown that failure to identify hazards, assess, and control the risks creates accident root causes that cause an environment that promotes high-risk behavior and high-risk conditions. These, as determined by Luck Factor 1, either result in a contact (exposure, impact, or exchange of energy) or a near-miss incident (Figure 6.1).

WARNINGS

These incidents have often been called near-miss incidents, narrow escapes, and warnings. Irrespective of their terminology, they are the real "safety in the shadows" according to S.L. Smith (1994) (p. 33).

We have been aware of incidents, near-miss incidents, and warnings for a number of years. Safety pioneer Heinrich (1959) introduced his 10 axioms of industrial safety back in the 1930s. His third axiom reads as follows:

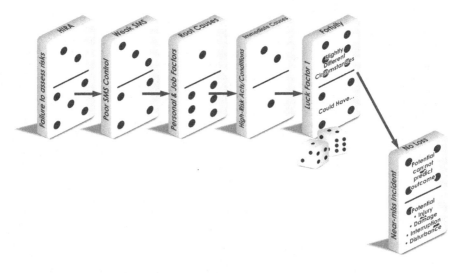

FIGURE 6.1 Near-miss incidents – potential for accidental loss.

 DOI: 10.1201/9781003385943-7

The person who suffers a disabling injury caused by an unsafe act, in the average case, had over 300 narrow escapes from serious injury as a result of committing the very same unsafe act. Likewise, people are exposed to mechanical hazards hundreds of times before they suffer injury.

(p. 21)

MISUSE OF TERMINOLOGY CREATES CONFUSION

As a trainee safety adviser, I was taught that an *accident* resulted in a loss and that an *incident* was a *near-miss incident, warning, close call or narrow escape.*

CONFUSING TERMINOLOGY

Current terminology creates confusion by all events being grouped under the same heading of *incident*. We had an undesired event at the office when the secretary dropped the coffeepot, which shattered, spraying hot coffee all over the rug. This was referred to as the "coffee incident." On the very same day, the newspaper reported there had been a malfunction at a nuclear power station, which was also referred to as an *incident*. To avoid confusion, it is recommended that we return to basics and redefine the concepts.

ACCIDENT AND NEAR-MISS INCIDENT

Bird and Germain (1992) define an accident as: "an undesired event that results in harm to people, damage to property or loss to process" (p. 18).

They define an *incident* as: "an undesired event, which under slightly different circumstances, could have resulted in harm to people, damage to property, or loss to process" (p. 20).

HIGH-RISK BEHAVIORS AND CONDITIONS

Many are not sure of the difference between high-risk behavior, a high-risk condition, an incident, or a near-miss incident or an accident.

HIGH-RISK BEHAVIOR OR CONDITION

High-risk behavior or *high-risk condition* is defined as: "a situation or act that has potential for loss." The word potential means the probable event is imminent, but no actions have yet taken place.

AN INCIDENT

OSHA defines an incident as: "an unplanned event that does not result in personal injury but may result in property damage or is worthy of recording."

They define a near-miss as: "an incident in which no property was damaged, and no personal injury was sustained, but where, given a slight shift in time or position, damage or injury easily could have occurred."

Safeopedia defines an incident as: "an incident, in the context of occupational health and safety, is an unintended event that disturbs normal operations."

CONFUSION

Other sources will give different definitions which lead to the confusion and results in safety professionals calling everything an accident/incident. To be sure the message is clear there needs to be a distinct difference between the two concepts and there must be no ambiguity or else more confusion will reign. This publication will refer to *near-miss incidents* as events with no loss, but with potential for loss, and *accidents* as undesired events that result in a loss.

NEAR-MISS INCIDENT

The minute there is an unplanned, uncontrolled flow of energy (with potential to cause loss), but no exposure, impact, or contact with this source of energy, arising from high-risk behaviors or conditions, then we experience a near-miss incident.

A near-miss incident requires the presence of a flow of energy, whereas a high-risk behavior or high-risk condition has not yet created an energy flow. Therefore, a high-risk situation is *not* a near-miss incident.

A near-miss incident usually refers to when there is a narrow escape or close call from injury. While there may be no injury (near-miss), the event may cause property or equipment damage or interruption. Since there is a loss, they are termed property damage or business interruption accidents.

AN ACCIDENT

An accident is an undesired event which results in a loss. The loss could be an injury, property damage, business interruption, or a combination of losses.

The following case studies are used to explain the differences between accident and a near-miss incident.

EXAMPLE OF AN ACCIDENT

An employee was descending the stairway from the first-floor office when his foot slipped on the second step from the bottom, causing him to fall. His elbow hit the bottom of the stair, causing bad bruising and a hairline fracture to the elbow. This is an accident, which has resulted in loss in the form of an injury.

EXAMPLE OF A NEAR-MISS INCIDENT

A worker was moving a heavy electric motor off of its mountings using a crowbar to slide the motor from its position. In tensioning the bar, it suddenly flew out of his hands, just missing the face of his colleague by a few inches. His colleague said that he felt the wind on his cheek as the flying bar flew through the air narrowly missing him. This event clearly fits the description of a near-miss incident with a flow of energy, but no contact with the energy (of the flying bar) and no loss.

MOST IMPORTANT STATISTIC

Perhaps the most important statistic in any health and safety management system (SMS) is the accident ratio showing the proportion of near-miss incidents, property damages, minor injuries, and serious injuries. Many authors and organizations have also compiled accident ratios but only between first aid cases, reportable injuries, and serious injuries. What has been left out in many cases is the bulk of the iceberg, the hidden part, the number of near-miss incidents or warnings that an organization gets before a major loss occurs.

ACCIDENT RATIOS

Heinrich was perhaps the first author to document the fact that for every serious injury there had been a number of incidents and near-miss incidents that could have given warning of the injury. His ratio, reproduced as Figure 6.2, proposed that for every 330 accidents there were 29 events that caused minor injury and 1 that caused serious injury. Three hundred caused neither injury nor damage and were termed "narrow escapes."

HEINRICH

Heinrich's third axiom of safety is that the person who suffers a disabling injury caused by an unsafe act, in the average case has had over 300 narrow escapes from serious injury as a result of committing the very same unsafe act. Likewise, persons are exposed to mechanical hazards hundreds of times before they suffer injury.

BIRD AND GERMAIN

In the 1966 accident ratio study, Bird and Germain changed the concept by proposing that for every 600 incidents (near-miss incidents), 30 property damage events occurred, 10 events resulted in minor injury, and 1 serious or major injury was experienced. This analysis was made of nearly 2 million accidents reported by ~300 participating companies employing 1.7 million employees (Figure 6.3).

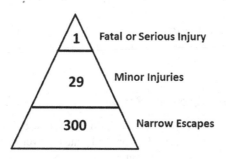

FIGURE 6.2 The Heinrich Accident Ratio (McGraw-Hill, 1959).

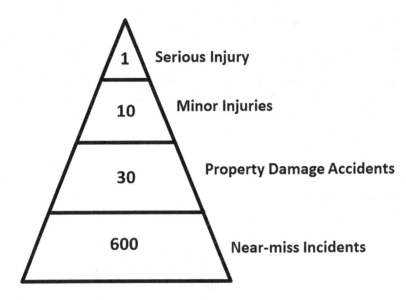

FIGURE 6.3   The Bird-Germain Accident Ratio.

As Bird and Germain (1992), report on the process involved in the study:

> Part of the study involved 4,000 hours of confidential interviews by trained supervisors
> on the occurrence of incidents that under slightly different circumstances could have
> resulted in injury or property damage. In referring to the 1:10:30:600 ratio, it should
> be remembered that this represents accidents and incidents reported and not the total
> number of accidents or incidents that actually occurred.
>
> *(pp. 20–21)*

THE TYE-PETERSON RATIO

In the United Kingdom during the years of 1974–1975, a study was conducted on
behalf of the British Safety Council. Based on a study of almost 1 million accidents
in British industry, the Tye-Peterson Ratio was deduced. The study was concluded
by stating that, "There are a great many more near-miss accidents than injury- or
damage-producing ones, but little is generally known about these" (British Safety
Council, *Safety Management*).

Figure 6.4 shows the Tye-Peterson Accident Ratio (1974/1975) showing that for
every 1 serious injury there were 3 minor injuries, 50 first aid, 80 damage-causing
accidents, and 400 near-miss incidents.

THE HEALTH AND SAFETY EXECUTIVE RATIO

In 1993 the Health and Safety Executive (HSE) of Great Britain came to a similar
conclusion with its accident ratio. It proposed that for every 1 serious or disabling
injury 11 minor injuries occurred and in excess of 441 property damages were
recorded.

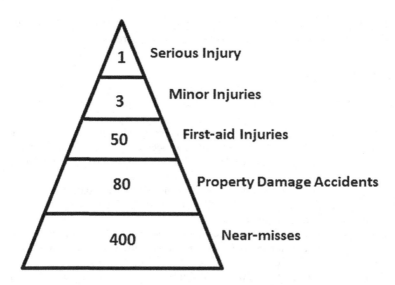

FIGURE 6.4 The Tye-Peterson accident ratio. (From the British Safety Council. 1974/1975 *Tye-Petersen theory.* With permission.) (From McKinnon, Ron C. 2012. *Safety Management, Near Miss Identification, Recognition, and Investigation.* (Model 2.2. Boca Raton: Taylor and Francis. With permission).

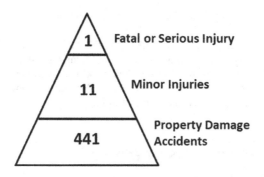

FIGURE 6.5 The Health and Safety Executive (HSE) accident ratio.

Figure 6.5 shows the HSE Ratio indicating for every 1 serious or disabling injury, 11 minor injuries occurred, and 441 property damage accidents occurred.

USA

A brief investigation as to the actual ratio of injuries to near-miss incidents was conducted at a large industry in the USA and the following ratio was determined. For every 1 disabling injury there were 15 reportable injuries, 4 first aid cases, 1 property damage, and 286 near-miss incidents. This ratio clearly showed that there was a correlation between reportable injuries and disabling injuries but also indicated the lack of reporting of property damage accidents. The near-miss incidents reported indicated a good near-miss incident program was in operation.

SOUTH AFRICAN RATIO

McKinnon (1992) in *Safety Management* magazine (September 1992), elaborated:

> A ratio such as this is truly invaluable since it makes prediction possible. As soon as too many near-miss incidents are reported in a specific area, it is a clear sign to the safety staff that a more serious accident is about to occur, and they can take preventive action.
>
> Bear in mind that the only difference between a near-miss incident and a serious accident is generally just a matter of luck. Take for example a brick falling off a platform and narrowly missing the head of a passerby.
>
> The accident occurred the minute the brick started to fall. The fact that it missed the passerby was not due to any safety institution but was purely fortuitous. While this is recorded as a near-miss incident, the implications are very serious.
>
> NOSA decided that statistics, which will enable them to construct a ratio, would be of great value to industry. "We want to determine whether there is a unique ratio in this country (South Africa) or whether Bird's Ratio of 1:10:30:600 was accurate."
>
> *(p. 12)*

TRENDS

Writing for *Occupational Hazards*, S.L. Smith (1994) says:

> Near-miss incident investigations war with the tradition of using an accident to trigger a thorough look at safety conditions and training. Williams suggested that if the purpose of safety programs is to prevent accidents, then tracking near-misses offers companies a better opportunity to lever their preventive efforts. Near-miss incidents can help firms pinpoint trouble areas and focus their safety training.
>
> *(p. 34)*

INJURIES – MEASUREMENTS OF FAILURE

Numerous industries, governments, organizations, and safety practitioners use the number and severity of injuries as the only measurement of safety. According to James Tye of the British Safety Council, these are measurements of failure. Measurements of injury and severity, as well as the costs thereof, are all reactive measurements and are not predictive. They are largely fortuitous anyway.

INJURY-FREE

Numerous organizations around the world have posters and banners depicting that their aim is to be injury-free. Numerous management-training programs also focus on injury prevention and tend to be specifically aimed at recordable injuries. Attention is given to that injury that requires reporting to the board of directors, the local safety and health enforcement agencies, or that injury that will ruin their "safety record."

POTENTIAL FOR INJURY

If one examines all the ratios and the research that has proven that there are numerous near-miss incidents before there is the occurrence of a severe injury, it can be deduced that it is impossible for a workplace to be injury-free unless it is accident-free. To be accident-free, it has to be near-miss-incident-free. To be near-miss-incident-free, there must be no high-risk behaviors and no high-risk conditions within the workplace. In the real world, it is almost impossible to eliminate all high-risk behaviors and high-risk conditions, but the near-miss incidents can give an indication of which high-risk behaviors and high-risk conditions have greater *potential* for causing the accident that in turn will cause the injury.

INJURIES ARE SYMPTOMS

Frank E. Bird Jr. convinced me that the key to safety success was, "If you look after the near-miss incidents, the accidents will look after themselves." One has to reduce the base of the triangle before the injuries that made up its peak can be reduced. Focusing only on injuries treats symptoms and not problems. After all, what sank the *Titanic*? The tip of the iceberg or the part hidden below the water line?

LUCK FACTOR

The only difference between an exchange of energy and a loss is Luck Factor 1. Krause (1997) supports this by saying,

> If we are lucky however, and experience only a near-miss, then that is just an incident.
> *(p. 292)*

S.L. Smith (1994) supports Krause's statement by saying:

> Most experts define a near-miss as any incident that could have led to damage to property or injury to employees but, for whatever reason, did not.
> *(p. 30)*

> In quoting Bruce Williams, S.L. Smith says that Williams calls near-miss incidents, "warning signs of bigger trouble to come."
> *(p. 30)*

Smith continues by quoting Fred A. Manuele, CSP, PE, President, Hazards Ltd., Arlington Heights, IL, as saying that,

> Investigations of near-miss incidents should receive higher priority in some cases than actual accidents.
> *(p. 33)*

NEAR-MISS INCIDENT REPORTING SYSTEM

One of the major weaknesses in numerous SMS is the fact that near-miss incidents are neither reported nor investigated. Unfortunately, only once an accident has occurred is an investigation conducted and corrective action taken.

As S.L. Smith (1994) says:

> But safety educator and author John V. Grimaldi, Ph.D., M.D.R., Calif., argues that the
> potential for severe exposure, even if nothing happened, is often a better indicator of
> the effectiveness of a safety program than injury rates.
>
> *(p. 33)*

FAILURE TO REPORT

Many organizations do attempt to get near-miss incidents reported by encouraging
employees to report them. After interviewing in excess of 300 employees as to why
they don't report near-miss incidents, the reasons given were that when an employee
reported a near-miss incident to his supervisor he was immediately challenged and
asked, "So what did you do about it?"

EXAMPLE

On one occasion, an employee reported to me that there was a 12-ft (4 m) excavation that
had been left totally unbarricaded. Another employee had almost driven his vehicle into
this ditch. When she reported this near-miss incident, her supervisor again challenged
her by saying, "So what did you do about it?" In most instances, the employee reporting
the near-miss incident does not have the necessary authority or the resources to take
action. In a number of the cases reported by the employees I interviewed, when they
did try to take action and things went wrong, they were accused of "operating without
authority," or doing something other than prescribed by their task description.

Often, doing something about a near-miss incident may involve giving instructions or advice to a more senior person and this immediately puts the employee in a
difficult position. As a result of this, employees have confided to me that they would
rather not report the near-miss incidents and not get involved.

PUNISHMENT

Another reason for employees not reporting near-miss incidents is that the employee,
if involved in the event, is often reprimanded for committing the high-risk behavior or
creating the high-risk condition. The employee is blamed and since he or she had made
the effort of reporting how their own behavior could have resulted in an injury or property damage accident; these efforts were rewarded with a rebuff. Fear of discipline and
repercussions is the other main barrier to the reporting of near-miss incidents.

OTHER BARRIERS TO REPORTING OF NEAR-MISS INCIDENTS

It is difficult to get employees to report near-miss incidents. The non-reporting may
be as a result of one or more of these common barriers:

- Employees don't know they are supposed to report near-miss incidents,
  after all, nothing happened.

- Employees don't know how to go about reporting the event.
- The training on near-miss reporting was insufficient, or the report methods are not clear.
- Employees are afraid of being reprimanded or disciplined for actions that led to the near-miss incident.
- Employees feel pressure from co-workers to keep quiet so that nobody gets into trouble, and nobody loses the safety bonus or spoils the "safety record."
- Workers are under pressure to maintain a clean incident, accident, and injury record so that the team will get the safety bonus or reward.
- They are new to the crew and want to make a good impression. Why make waves and stand out in the crowd all for nothing?
- The work safety culture says, "Nothing happened, no one was hurt so don't make a big deal out of it."
- Co-workers view the near-miss incident with humor instead of seeing the hazard. If everyone is laughing, how serious could it be?
- Last time they tried to talk to the supervisor about a near-miss incident they were belittled or disregarded.
- Workers find it too much trouble filling out those forms and they have no time for paperwork.
- Employees, in general, dislike paperwork.
- "We tried other safety things in the past and they didn't work."
- Near-miss incident reporting is not encouraged by the organization.
- Language barriers.

There are other barriers to be overcome. These include:

- Potential recriminations for reporting such as the fear of disciplinary action, fear of peer pressure, teasing, and concern for being involved in an investigation.
- Motivational issues such as the lack of incentive and management discouraging near-miss reports.
- Lack of management commitment reflected by sporadic emphasis, and management fear of liability.
- Individual confusion as to what constitutes a near-miss and how it should be reported.

Anonymous Reporting

The first and most important aspect of a near-miss incident reporting system is that it must be a "no names, no pack drill," or anonymous reporting system. Management should be interested in *what* happened and not *who* messed up. As all these accident ratios indicate, there are always plenty of mess-ups before a serious injury occurs, and as long as people are working in an industrial environment, mistakes will be made. The objective of a near-miss incident reporting system is to find out what actions can be taken to prevent further undesired events before they result in a loss.

Near-Miss Incident Reporting Form

The next step in a near-miss incident reporting system is to devise a simple, yet effective near-miss incident reporting form. Most employees and supervisors dislike paperwork, especially safety-related forms and questionnaires. As a result of this, a short, brief, near-miss incident reporting form will reduce the resistance to this "safety paperwork." Electronic reporting systems eliminate paperwork and encourage quick and instant reporting.

The reporting form should contain the following:

- person's name (as an optional)
- the date of the event
- the location of the event
- a brief risk assessment of the event
- a description of what happened
- action taken (if applicable) or recommended

The forms should be readily available throughout the workplace. Ideally, they should be bound in a pocket-size booklet so that the employee always has forms available. If possible, the book should be carbonized so that the employee retains a copy of the near-miss incident reported. Workers can then monitor the follow-ups at their own pace.

Electronic Reporting Systems

Cell phone apps have been developed that allow the reporting of near-miss incidents and other high-risk situations in the workplace. These should be incorporated into the SMS as formal reporting systems.

Success

The introduction of such a *no blame* near-miss incident reporting system facilitated by pocket-size reporting booklets improved near-miss incident reporting from 10 near-miss incidents per month to ~300 per month at one organization.

Training

Correct training on how to use the system also boosted its usefulness. Previously, at one organization, forms were distributed haphazardly, and no training was given. A two-hour training session describing the importance of near-miss incident reporting, how to use the form, and how to conduct a risk assessment of the near-miss incident potential was met with tremendous success and resulted in more than 300 near-miss incidents being reported in the first month (Figure 6.6).

Flow Process

Every near-miss incident reporting system should have a flow process showing what happens to the near-miss incident form or report that has been handed in for

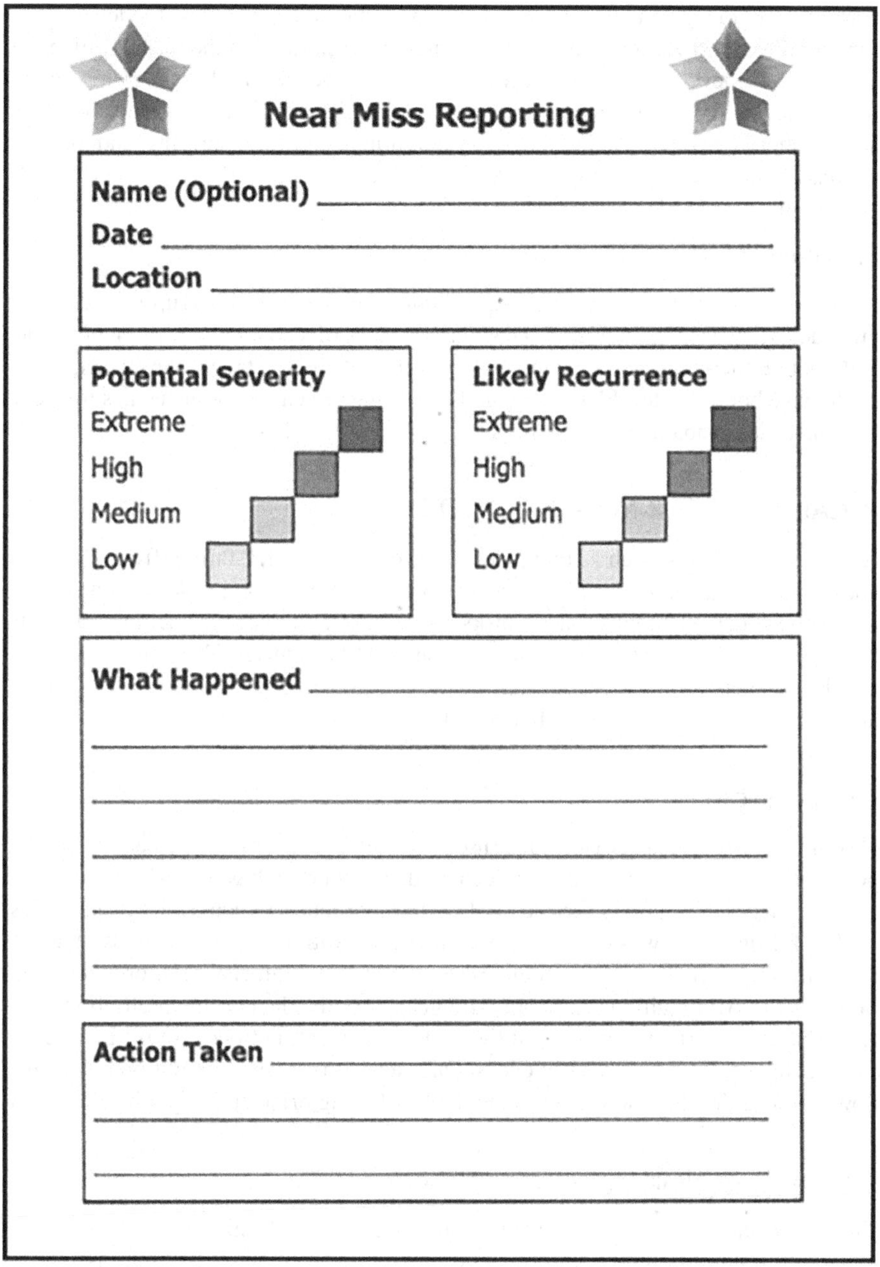

Near Miss Reporting

Name (Optional) ____________

Date ____________

Location ____________

Potential Severity

Extreme

High

Medium

Low

Likely Recurrence

Extreme

High

Medium

Low

What Happened ____________

Action Taken ____________

FIGURE 6.6 An example of a near-miss incident reporting form incorporating a simple risk assessment matrix.

corrective action. The near-miss incident report should have ranked the level of risk (potential for loss) via the mini risk assessment incorporated in the forms, which will determine the level of action that the event will receive. The higher the risk potential and severity is ranked, the higher the priority of rectifying the hazard. The completed actions should be channeled to the safety committees for their scrutiny, and, once all actions are completed, the form is filed for future reference.

PREDICTIVE

Accidents that result in injury, property damage, or business interruptions are often preceded by near-miss incidents. The consequences of a contact may often be hidden and, though there is contact, under slightly different circumstances the consequences could have been greater. Identifying and rectifying the causes of near-miss incidents eliminates all probable consequences.

RECALLING NEAR-MISS INCIDENTS

A near-miss incident such as the undesired event of the brick falling from a scaffold above, but with no damage or consequence, would form one of the plenty of near-miss incidents, which eventually lead to property damage and injury-producing accidents. Near-miss incidents are often the foundation of major injuries. They are the same as accidents except for the missing phase of the exposure, impact, or exchange of energy above the threshold limit of the body or structure.

INCIDENT RECALL

Near-miss incident recall is a technique used to recall these near-miss incidents which had potential to cause loss but were never identified nor reported. It gets employees thinking about instances they have experienced where the consequences could have been serious.

The falling brick was caused by a high-risk condition. Most near-miss incidents that will be recalled are either high-risk conditions or high-risk acts, or a combination of both. Although these are not pure near-miss incidents as per definition, their reporting is important. They cause the undesired event, but because of the circumstances do not result in any visible loss. Once a high-risk act or condition results in a flow of energy we have a near-miss incident. A falling brick creates a flow of energy.

IMPORTANCE OF NEAR-MISS INCIDENT RECALL

The importance of near-miss incident recall cannot be emphasized enough. Accidents that cause injury, damage, and production losses are recognized as being important and are reported and investigated. Many near-miss incidents are ignored and not reported or acted on.

Near-miss incident recall reveals the events that could have injured someone or damaged something. Near-miss incident recall is the ideal way of getting near-miss incidents reported, so they can be investigated, and positive steps can be taken to eliminate the root causes.

Near-miss incident recall can also offer a guide as to what corrective actions are needed concerning the high-risk acts and conditions that are being reported. It can also help to indicate certain trends.

Example

During an incident recall session one participant recalled how he was travelling to work on the freeway when suddenly a three-seater couch landed up in front of him on the road. He managed to change lanes and avoid a collision. The recall reminded all to secure any loads on a pickup or truck so that they do not fall off while travelling on a busy freeway.

BENEFITS OF ACCIDENT RECALL

Accident recall revisits previous work injuries or occupational diseases or damage accidents, which have already been investigated and analyzed. It is also a proactive safety measure, based on post-contact events, and it acts as a reminder of accidents that happened in the past. It also offers a guide to ensure that follow-up is still in force as a result of previous accident investigation recommendations.

Accident recall gets active participation from and shares valuable experience with the team. It helps to motivate as it indicates that past accidents are not merely forgotten, and rated unimportant, but are recalled frequently and constant vigilance is thus maintained. Accident recall is ideal for induction training and also for making the new worker aware of the type of accidents that have occurred in the past and which may occur in the future.

BENEFITS OF NEAR-MISS INCIDENT RECALL

Near-miss incident recall gets employees involved in the safety process. It encourages them to contribute to the safety of the workplace by getting them to recall near-miss events that happened either at work, at home, or during recreation activities. This sharing of "what could have happened" is a learning experience for all participants. Employees are exposed to many risks in the workplace and the best way of gaining insight into near-miss experiences is via formal incident recall sessions.

PRE-CONTACT AND POST-CONTACT ACTIVITIES

Near-miss incident recall is a pre-contact activity. It allows the recall of near-miss incidents that have not yet caused harm or damage and endeavors to identify the root causes and rectify them before any contact takes place.

Accident recall is a post-contact safety activity, as a loss must have occurred before the accident could be recalled, reminding workers of its details.

DISCIPLINE

Employees will be hesitant to recall near-miss incidents if the recall ends up in disciplinary measures. A near-miss incident recall session should be treated as confidential, and the information disclosed should also be handled with discretion. No disciplinary measures should take place. Finger-pointing and punishing people for recalling near-miss incidents will stop employees participating in the near-miss incident recall sessions. It will also inhibit employees from using the system to report near-miss incidents. Where possible, the near-miss incident recall should be a fact-finding session and not a blame fixing or witch hunt activity.

METHODS OF RECALL

There are various methods of conducting both near-miss incident and accident recall sessions. The two main methods are the formal and informal recall sessions.

Formal Recall

Formal recall can be tabled as an agenda item at monthly health and safety committee meetings or during toolbox talks. At the end of health and safety training sessions a five-minute recall session could be held and near-miss incidents could be recalled by the employees attending the training session.

Informal Recall

Informal recall sessions can be held on a person-to-person basis during the normal workday. A near-miss incident recall form could also be available, and anybody could then fill in the form and report a near-miss incident. This report could remain anonymous and the person reporting the near-miss incident need not name the persons involved.

Major Loss Briefing

A major loss briefing is an accident recall session that takes place after a major accidental loss. This major loss could be a fatal or permanent injury, serious property damage, or a combination of both.

A description of the accident is presented at the meeting or gathering, and copies of the event distributed to everybody in the plant to keep them informed. This description is also pinned up on the various notice boards and circulated on the company intranet so that complete exposure is achieved, and awareness is created among the employees.

This major loss briefing is an ideal way to make the workforce aware of the causes of the accidents, and also the remedial steps that have been taken to prevent a recurrence.

Safety Stand-Down

After a fatal accident many organizations down tools and have a safety stand-down. This is when work stops; employees are grouped in various halls or lecture rooms and are addressed by management and union leaders. The discussion is almost a last ditch stand in an effort to stop high-risk situations that have already had dire consequences. It is used as a major "wake up call" to all, that what happened is unacceptable and that the situation must improve from now on.

The stand-down is a company-wide accident recall session. The details of the event, normally a fatal accident, are recalled, analyzed, and a line is drawn in the sand with management demanding an improvement in safety cultures, habits, and performance from that point on. Regrettably the stand-downs are normally held after the event rather than before the event.

Safety Stand-Down Based on Near-Miss Incidents

Below follows a write up of a safety stand-down that was held involving some 350 plant employees.

Safety Stand-Down Day at Plant Operations
Under the leadership of the plant manager, the entire development section of plant operations, consisting of some 350 employees participated in an 8-hour safety stand-down. The stand-down covered all employees within the division and took place on the A shift on Tuesday, 4 June.
The very intense program was initiated as a result of serious injuries being experienced in the development section and also as a result of the types and high potential of near misses being reported from the divisions.

PREDICTIVE SAFETY MANAGEMENT

Reporting, recording, investigating, and taking corrective actions on near-miss incidents is predictive safety management. Near-miss incidents are warnings of a failure in the system that is about to culminate in a major loss. In many cases, near-miss incidents are more important than accidents. The most important words in safety are, "It's not *what* happened, but what *could have* happened."

Investigating and rectifying causes of high-potential near-miss incidents will eventually lead to a reduction in the number of accidents and their resulting injuries.

As W.E. Tarrants (1980) said,

It is appropriate to consider possible means for estimating prospective accident experience through some direct measure of *accident propensity*. To be useful for this purpose, it is desirable, but not necessary, to be able to transform measures of propensity directly into estimates of future accident experience, or at least insofar as the objective is to make relative rather than absolute assessment.

(p. 187)

As is proposed by the CECAL loss causation sequence, the only difference between an accident and a near-miss incident is the outcome, which is largely fortuitous.

RISK ASSESSMENT

The risk assessment of reported near-miss incidents is important. In some instances, safety practitioners recommended to management that *all near-miss incidents* be investigated. This led to a tremendous workload and a waste of time and effort. Near-miss incidents are important, but those with high loss-potential severity and high probability of recurrence should receive priority when it comes to investigation.

Risk Matrix

An example of a simple risk matrix such as that produced in Figure 6.7 can be included on the incident reporting form.

The person reporting the near-miss incident merely has to check a block as to the loss potential severity of the incident, (or what could have happened), as well as the probability of the event happening in the future.

A risk ranking of the near-miss incidents can then be established. Some near-miss incidents can be rectified immediately, and the action taken can be entered on the form.

Often a ranking of (1–5) indicates a risk that can be managed and rectified immediately. Events that fall in the gray (4–10) area of the matrix should be investigated. Events in the (9–15) range should get higher priority. Those that fall in the dark gray (16–20) area should receive the same prominence and attention as a fully-fledged accident investigation of an injury, property loss, or business disruption. Those that rank in the black (25) should be treated urgently as they have the highest potential for accidental loss.

S.L. Smith (1994) also references Grimaldi:

> Grimaldi counsels safety professionals to rate near-misses in order of their potential for severe exposure and consequence and tackle the big ones first.
>
> *(p. 34)*

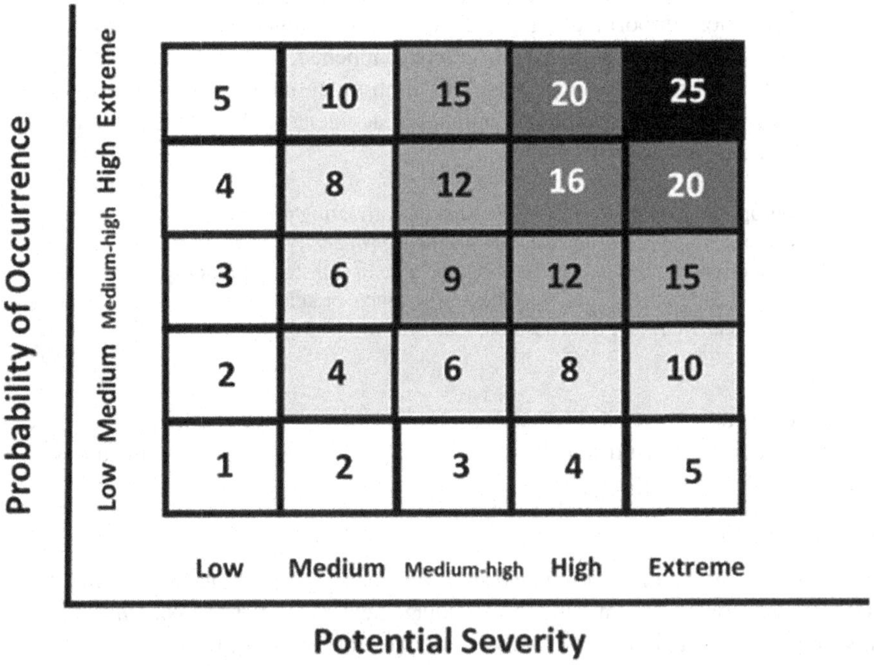

FIGURE 6.7 A risk matrix.

PRO-ACTIVE SAFETY CONTROLS

S.L. Smith continues:

> But a growing number of safety experts now agree that near-miss incidents may provide a better measure of the effectiveness of a safety program than tracking injuries and that their reporting and investigation are critical in preventing injuries and fatalities from occurring. Most experts define a near-miss as any incident which could have led to damage to property or injury to employee but for whatever reason did not.
>
> *(p. 33)*

Near-miss incident trends can be calculated and used to initiate pro-active SMS control measures.

RESULTS

Figure 6.8 shows near-miss incidents reported per month as well as the actions taken. These actions are pre-contact, pro-active safety activities initiated to prevent a recurrence of these warnings, which, under slightly different circumstances, could have resulted in loss. The rectification of some hazards revealed by the near-miss incident

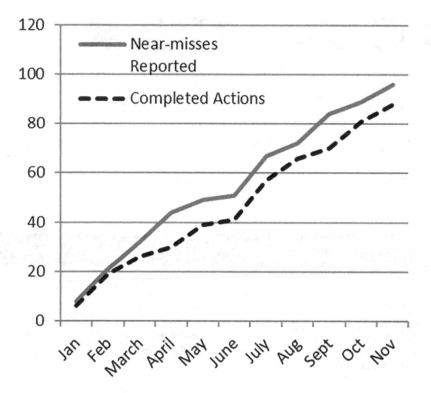

FIGURE 6.8   Near-miss incident reporting and corrective actions tracking system.

reporting system may only be rectified in the long term. Others may be rectified in a short time.

RISK MATRIX ANALYSIS

Using a near-miss incident reporting system, a risk matrix analysis on potential severity and frequency was compiled and is reproduced as Figure 6.9. This table shows that the majority of the near-miss incidents had *medium* loss-severity potential and probability of recurrence was *low*. However, nearly 25% of the events had high potential for loss severity and 6% had a high probability of recurrence.

POTENTIAL EXPOSURE, IMPACT, OR ENERGY EXCHANGE TYPE

The analysis also analyzes the potential exposure, impact, or energy exchange that could have occurred. It shows that 30% of the contact types would have been *struck by*, 20% would have been *slips*, 18% *foreign objects*, and 10% *falls to lower levels*.

Figure 6.10 shows an analysis of the probable exposure, impact, or energy exchange type that could have occurred under different circumstances derived from a near-miss incident reporting system.

Severity	Number	Percentage	Frequency	Number	Percentage
Low	66	36%	Low	105	57%
Medium	75	41%	Medium	68	37%
High	44	24%	High	12	6%

FIGURE 6.9   Potential severity and frequency analysis.

Potential Energy Exchange Type	% Events	Potential Energy Exchange Type	% Events
Struck against	6%	Fall lower level	10%
Struck by	30%	Overexertion	1%
Caught in	1%	Contact electricity	2%
Caught on	4%	Acids	1%
Caught between	6%	Toxic substances	2%
Slip	20%	Foreign objects	18%
Fall same level	1%		

FIGURE 6.10   Potential energy exchanges.

Using this information, management can then almost predict the severity, the frequency of recurrence, and the type of the next contact, or energy exchange. This compilation of narrow escapes serves as warnings as to what could happen if it were not for Luck Factor 1.

ONGOING ANALYSIS

The next month's analysis of the same reporting system resulted in 215 near-miss incidents reported with 184 actions being taken. The analysis of the month of September also showed that 3% of all near-miss incidents had high loss-potential severity and 5% of the incidents had a high probability of recurrence.

The potential contact type changed to potential *struck by* – 28%, followed by potential *striking against* – 19% and then *caught on* – 10%.

FEEDBACK

One of the most important steps in the near-miss incident reporting system is giving feedback to the employees who report near-miss incidents. The most obvious form of feedback is the correcting of the situations that have been reported and the taking of actions to prevent the symptoms from recurring.

Some organizations have a monthly lucky draw for near-miss incident reporting recognition. All the members of the division who have reported five or more near-miss incidents put their names in a hat. A winner is drawn. A small prize is then presented to this person to thank him or her for the contribution. Tremendous enthusiasm can be generated by a simple competition.

As is stated by many, the work of any good manager should be to reduce the "plenty" of near-miss incidents.

INJURIES VIS-À-VIS NEAR-MISS INCIDENTS

By constantly eroding the base of the iceberg or accident ratio triangle, successes will eventually start to impact the top of the triangle, i.e., the minor and serious injuries.

An interesting example of this was a study carried out at a large refining works. A near-miss incident reporting system was encouraged, and from a mere 50 to 60 near-miss incidents reported in a month, this figure jumped to more than a hundred a month within a few weeks.

In superimposing the reduction of injuries on the number of near-miss incidents reported, it was found that as the near-miss incidents reported increased, so the number of injuries fell. Near-miss incidents reported rose from 75 in January to 166 in February and 231 in March. The injuries reduced from 5 in January to 4 in February and 3 in March. There was a further reduction down to one injury in the month of April. As the number of near misses reported fell, the injury rate once again continued to climb (Figure 6.11).

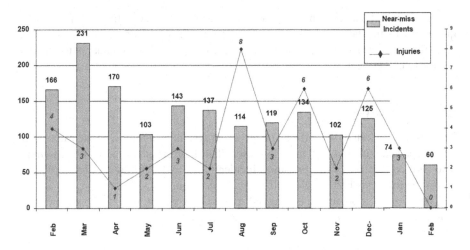

FIGURE 6.11   Near-miss incidents vis-à-vis injuries.

SUMMARY

Near-miss incidents are warnings of a failure in the management system, and they should be treated with the same vigor as an accident. High potential near-miss incidents should also be investigated, and the root causes identified and eliminated; this will reduce the number and frequency of accidents.

7 Potential for Loss

The sequence of events described so far have encountered Luck Factor 1 and have resulted in either an exposure, impact, or contact with a source of energy or a near-miss incident. If the event is a near-miss incident, then the potential of that event to cause loss under slightly different circumstances must be assessed (Figure 7.1).

TREATING NEAR-MISS INCIDENTS SERIOUSLY

Traditionally, most near-miss incidents are ignored because "nothing happened," or so it seems. In referring to a number of safety pioneers' accident ratios, it is obvious that there are numerous near-miss incidents that occur for every serious injury experienced. The actual numbers are not that significant, but the fact that there is a correlation is important. It would then seem logical that all these near-miss incidents should be treated as seriously as loss-producing accidents if they have high potential for loss.

MISLEADING

Some safety practitioners often advise managers to investigate every single near-miss incident. They believe that each near-miss incident could end up as an injury. This is misleading. Due to manpower and cost constraints, it is virtually impossible, as well as totally impracticable, to investigate *all* near-miss incidents.

POTENTIAL

In reviewing the near-miss events, the keyword is *potential*. What did the event have the capacity to do under slightly different circumstances? Near-miss incidents are

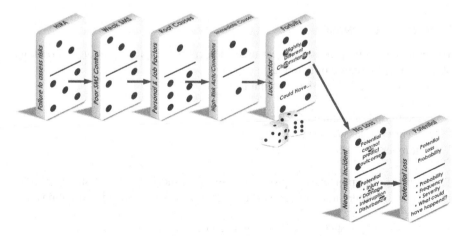

FIGURE 7.1 Near-miss incidents have potential to cause loss.

DOI: 10.1201/9781003385943-8

early warnings of potential problems. The potential (capacity for accidental loss) is what should determine which of these near-miss incidents should be investigated, as well as the level of the investigation required.

POTENTIAL HAZARDS

Safety practitioners have often debated the expression "potential hazard" and have concluded that there is no such thing as a *potential* hazard.

They argue that a hazard is a hazard is a hazard. They are correct. The term *potential* hazard really refers to a hazard that has the *potential* to cause harm. Once again, the keyword being *potential* or having the capacity to cause harm.

LOSS POTENTIAL

Contemporary thinking is that the existence of an injury or property damage loss is no longer a necessary condition for reviewing accident performance. It is now possible to identify and examine accident problems before they occur instead of "after the fact" in terms of their injury-producing or property-damaging consequences. This allows management to concentrate on measurement of loss potential or near miss incidents and remove the necessity of relying on measurement techniques based on the probabilistic, fortuitous, rare-event, injurious accident.

Boylston (1990) refers to near-miss incidents as potential problems and also quotes the luck factors:

> Failure by an organization to recognize, evaluate, and implement controls for early warnings of potential problems usually results in a system of reactive approaches. Consequently, there is little if any way to control the magnitude of the problem. Such organizations are "lucky" or "unlucky," depending on the situation. This is no way to manage an organization.
>
> *(p. 103)*

RANKING THE POTENTIAL

The most important aspect of a near-miss incident is the quantification and ranking of the incident's degree of potential.

The potential could be the potential for:

1. likelihood of occurrence (probability)
2. severity of loss (severity)
3. number of, and how often (frequency)

These are the common terms of probability, severity, and frequency. A simple method of ranking the potential of a near-miss incident is to ask the following questions:

- What is the probability of this event occurring?
- If this event occurs, how bad will the consequences be?

- If the event occurs, how often will it be repeated and how many people are exposed?

Often only two criteria are considered. They are the probability of occurrence and possible severity of that loss.

SAFETY SOLUTION

It is strongly felt that the solution to safety problems lies within those near-miss incidents with high potential, as they are accidents that the organization has not yet experienced. The loss causation sequence has been triggered, but, due to Luck Factor 1, has ended in a warning, a near-miss incident. All management need do, is identify the potential of the event. If the potential is high in terms of probability and severity, treat the near-miss incident as if something *had* happened and institute appropriate control measures.

CRYSTAL BALL

A very simple method of assessing the potential of each near-miss incident is to gaze into an imaginary crystal ball and quote the magic words "it's not *what* happened, it's what *could have* happened." This very simple potential assessment technique will identify those near-miss incidents with the greatest potential for loss and which should be treated with the same urgency as loss-producing accidents.

POTENTIAL LOSS AND SEVERITY RANKING SEQUENCE

A very simple method of receiving, assessing, and analyzing the potential of near misses is as follows:

1. Train and encourage employees to report all near-miss incidents, irrespective of their potential.
2. There should be anonymity of reporting.
3. Assure them that there will be no repercussions and that the system is a "no names, no pack drill" exercise.
4. Issue a simplified near-miss incident report form, that includes a risk matrix, or develop an electronic reporting system.
5. Commend employees on submitting near-miss incident reports.
6. Using a simplified risk matrix system, ideally, have the reporter rank the probability of recurrence and potential severity of each near-miss incident reported.

Figure 7.2 is such a matrix, rating the probability of recurrence and potential severity from low to high. As discussed in Chapter 6, often a ranking of (1–5) indicates a risk than can be managed and rectified immediately. Events that fall in the gray (4–10) area of the matrix should be investigated. The same accident investigation report and process should be used, and the same diligence applied even though there was no loss. The potential for loss is the guiding factor in this instance.

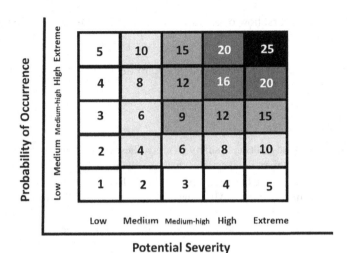

Potential Severity

FIGURE 7.2   An example of a risk matrix.

Events in the (9–15) range should get higher priority. Those that fall in the dark gray (16–20) area should receive the same prominence and urgent attention as a fully-fledged accident investigation of an injury, property loss, or business disruption. Those that rank in the black (25) should be treated with upmost urgency as they have the highest potential for accidental loss.

Normal Expected Loss

In ranking the potential of near-miss incidents one could picture the worst-case scenario, which would raise the potential of most near-miss events. It is advisable to use the *normal expected loss* approach by asking the question, "What is the probability of recurrence and the loss-potential severity under normal circumstances?" Normal expected loss means being based on, or using good judgment, and therefore being fair and practical. Extreme outcomes such as worst-case scenarios should not be predicted.

Examples

These reports are from a processing plant that used a simplified risk matrix as in Figure 7.3.

1. In removing the liners from a mill, an employee left a 3-ft (900 cm) pry-bar lying in the bottom of the mill casing. To remove the liners, the nuts are removed, employees withdraw to a safe distance, and the mill is rotated, allowing the liners to fall to the bottom of the mill. On this occasion, the liner fell on the pry-bar which had been left in the mill, sending it flying through the air like an arrow ~4 ft (1.2 m) from the ground. Fortunately, it narrowly missed the supervisor as it flew past him at high speed. The

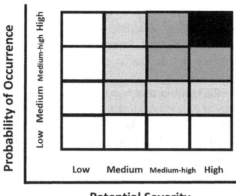

FIGURE 7.3 A simplified risk matrix.

potential of this near-miss incident was ranked as extremely high, as was the probability of recurrence. Witnesses agree that under slightly different circumstances, the pry-bar could have hit the supervisor causing serious injury or death.

2. While hoisting a ramp, a cable broke and the ramp fell, narrowly missing a worker who jumped out of the way. The ramp weighed in excess of 1 ton (900 kg) and falling from a height of 12 ft (3.6 m) had the potential to cause serious injury to the worker. The probability of recurrence was rated high, as the raising of the ramp was a daily activity. This was also investigated as a property damage accident.

3. An employee was walking in a walkway where a piece of grating was missing. He nearly fell into a ditch as a result of this hazard. This near-miss incident was ranked as low potential as the ditch was only 1 ft (30 cm) deep. The probability of recurrence was also ranked low as this was an isolated occurrence of the grid being removed.

4. An employee started work on an electric motor and received a slight electric shock. The potential of this near-miss incident was ranked as extremely high and the probability of recurrence as medium-high. An electric shock situation has high potential to cause death, and this is one instance with high potential that deserves thorough investigation and the implementation of remedial measures. The difference between a slight electric shock and electrocution is once again fortuitous.

5. While lifting a 300-lb (130 kg) battery, the battery cable broke and the battery fell 4 ft (1.2 m), just missing an employee's feet. Considering the size of the battery and the distance falling, there was obviously potential for severe injury to the person's feet, had he not jumped clear. This near-miss incident was ranked as medium-high on the probability of recurrence as this was a daily activity, and loss potential severity was ranked as medium-high. This would rank the potential of the near-miss incident as sufficient to warrant a further investigation.

Classic Example

An interesting near-miss incident was reported by a divisional manager of a storage facility. The report was as follows: Supervisor was assisting truck driver to load vehicle with scrap batteries. A truck battery fell off a hand truck onto his foot. He was wearing safety shoes.

This was ranked by the reporter as a minor event with low injury potential. Upon further investigation it was pointed out that the event had high potential for injury as the batteries were heavy items and could inflict severe injury if one fell onto a person's foot. The fact that the supervisor was wearing safety shoes does not reduce the potential of the event. Safety shoes do reduce the energy exchange and prevent injury to the foot, but do not diminish the potential for injury that the falling battery had.

SUMMARY

Near-miss incidents are warnings that there is a weakness in the SMS, or that the health and safety management system is flawed in some respect. Not all near-miss incidents, however, need immediate and in-depth reactions as some near-miss incidents have less potential than others do.

Assessing the potential of a near-miss incident gives a clear indication of the event's potential so that investigations can be prioritized. Using a simple risk matrix, each near-miss's potential can be ranked, and consequent investigations and follow-up actions prioritized. Near-miss incidents should be treated as friendly warnings. Friendly warnings that have high potential offer a clear indication of what *could* happen under slightly different circumstances, which are normally beyond our control. Management should take heed of these high potential near-miss incidents and institute controls before a loss occurs. Should an exposure, impact or contact with a source of energy take place, the consequence of that contact is fortuitous.

8 Exposure, Impact, or Exchange of Energy

Once a high-risk behavior or a high-risk condition is imminent, Luck Factor 1 determines whether there will be an exposure, impact, or exchange of energy. This exchange is in the form of a contact with a substance or source of energy greater than the threshold limit of the body or article. The high-risk behavior or high-risk condition may result in a near-miss incident, where, although there was potential, there was no exposure, impact, or contact with a source of energy (Figure 8.1).

The exposure, impact or contact, or exchange of energy is the part of the CECAL sequence that is most closely associated with the loss. The exposure, impact, or exchange of energy is what injures, damages, pollutes, or interrupts the business process.

INJURY

The National Safety Council's definition of an injury, given in Injury Facts 2020, further describes this exchange of energy as follows: "an injury is physical harm or damage to the body resulting from an exchange, usually acute, of mechanical,

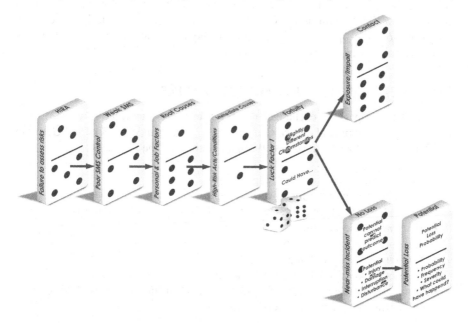

FIGURE 8.1 The domino depicting the exposure, impact, or exchange of energy in the accident sequence.

DOI: 10.1201/9781003385943-9

chemical, thermal, or other environmental energy that exceeds the body's tolerance" (NSC Website, 2022).

The accident is the entire undesired event constituting the CECAL sequence. The contact part (domino) of the sequence is what causes the actual loss.

CONFUSION

Many refer to the contact phase of the sequence as the accident. The definition given by Heinrich (1959) is that: "an accident is an unplanned, uncontrolled event in which the action or reaction of an object, substance, person or radiation results in personal injury or the probability thereof" (p. 23).

In explaining his domino theory of accident causation, Heinrich's 4th domino is entitled the accident and is explained as:

> Events such as falls of persons, striking of persons by flying objects, etc., are typical accidents that cause injury.
>
> *(p. 23)*

To clear any confusion, the *accident* referred to by Heinrich is the exposure, impact, or exchange of energy and the contact, which does the harm.

The National Safety Council glossary still defines an accident as: "that occurrence in a sequence of events which usually produces unintended injury, death or property damage" (NSC Website, 2022).

This definition, as well as Heinrich's, leads one to believe that the exchange of energy is the accident and not the entire sequence of events. In referring to the Bird and Germain (1992) definition, the accident is defined as: "the undesired event that results in a loss to people, etc." The event is the sequence that leads up to and includes the exposure, impact, or exchange of energy.

ENERGY EXCHANGE – NOT ACCIDENT TYPES

The exposure, impact, or exchanges of energy are also (incorrectly) referred to as accident types. A more apt description would be *energy transfers*. Bird and Germain (1992) (from ANSI ZI6.2–1962, rev.1969) quote these as the following:

- Struck against (running or bumping into)
- Struck by (hit by moving object)
- Fall to lower level (either the body falls, or the object falls and hits the body)
- Fall on same level (slips and fall, tip over)
- Caught in (pinch and nip points)
- Caught on (snagged, hung)
- Caught between (crushed or amputated)
- Contact with (electricity, heat, cold, radiation, caustics, toxics, and noise)
- Over stress/over exertion/overload (p. 26)

The terms *types of exposure, impact,* or *contact* or *energy exchange* are far more accurate descriptions than accident types. In the CECAL sequence, a distinction will be made between the accident itself and the specific type of exposure, impact, energy exchange, or energy transfers.

Loss Causing

As stated, the exposure, impact, or energy exchange in the sequence of events is what causes the loss.

In one accident investigation it was found that the exposures, impacts, and contacts that took place were:

1. Overexertion – the weight and movement on top of the utility pole was greater than the compaction of the base of the pole, causing it to topple.
2. Fall to below – the person and the pole fell to the ground.
3. Struck by – the injured person was struck by the pole when it hit the ground and bounced back hitting him.
4. Fall to below – after being hit by the rebounding pole, the worker then fell onto the ground.

The above is a contact analysis of the various impacts and exchanges of energy that took place during the accident and that caused injury, damage, and business interruption.

EXPOSURE, IMPACT, AND CONTACT EXPLAINED

Heinrich et al. (1969), in explaining the Bird/Germain loss causation model, describe contact as follows:

> The word "contact" appears on the domino at this point in the sequence because an ever-increasing number of researchers and safety leaders around the world look at the accident as a "contact" with a source of energy (electrical, chemical, kinetic, thermal, ionizing, radiation, etc.) above the threshold limit of the body or structure, or "contact" with a substance that interferes with normal body process.
>
> *(p. 27)*

In further explaining how Bird and Germain intended the contact domino to fit into the sequence, they continued:

> This point in the sequence is also referred to as the contact stage, and applications of the principles of deflection, dilution, reinforcement, surface modification, segregation, barricading, protection, absorption, and shielding are examples of counter measures frequently used as loss control tools here.
>
> *(p. 27)*

Heinrich et al. (1969) explain Dr. Michael Zabetakis's theory on unplanned transfer or release of energy as follows:

> Most accidents are actually caused by the unplanned or unwanted release of excessive amounts of energy (mechanical, electrical, chemical, thermal, ionizing radiation) or of hazardous materials (such as carbon monoxide, carbon dioxide, hydrogen sulfide,

methane, and water). However, with few exceptions, these releases are in turn caused by unsafe acts and unsafe conditions. That is, an unsafe act or an unsafe condition may trigger the release of large amounts of energy or of hazardous material which in turn causes the accident.

> *(p. 32) (Dr. Zabetakis's use of the word accident here should read "loss.")*

PETERSEN MODEL

The Petersen Accident-Incident Causation Model, Heinrich et al. (1969), shows the accident immediately preceding the injury or loss (p. 49). The CECAL Model puts the contact immediately before a loss. The Ball Model quoted by Heinrich et al. (1969) confirms that:

> Ball's theory is that all accidents are caused by hazards and that all hazards involve energy, either due to involvement with destructive energy sources or due to a lack of critical energy needs. The concept of energy release is a necessary part of the accident-causation process.

> *(p. 53)*

CONTACT RISK ASSESSMENT

As most safety researchers seem to agree that there is some form of exposure, impact, or energy release that precedes the injury, damage, or interruption, the consequences of certain types of energy exchange can be predicted by applying a simple risk assessment.

The risk assessment of the possible contact would be the answer to the question, "If it happens, how bad would it be?" To estimate the frequency of the energy exchange, ask, "What is the probability of recurrence of this exchange of energy?" Ranked on a scale of low, medium, and high, a high-high rating would indicate greatest potential loss from that particular contact.

AGENCY

The agency is the piece of equipment or object closest associated with the loss and is defined by Simonds and Grimaldi (1963) as: "the substance, object, radiation, or person most closely associated with the accident's occurrence" (p. 178).

They list a few general groupings of agencies, which include:

- animals
- boilers and pressure vessels
- chemicals
- conveyors
- dusts
- electrical apparatus
- elevators
- hand tools
- highly flammable and hot substances

- hoisting apparatus
- machines
- mechanical power transmission equipment
- prime movers and pumps
- radiation and radiating substances
- working surfaces and miscellaneous (p. 197)

The National Safety Council (1993) defines the agency or agent as: "the principal objects such as tools, machine or equipment involved in the accident and is usually the object inflicting injury or property damage" (p. 111).

The agency is therefore a key factor in the exposure, impact, or exchange of energy. It is the agent that is responsible for the energy transfer to the recipient who consequently suffers a loss in the form of injury or illness. The loss could be damage or interruption.

Agency Part

The agency part is that part or area of an agency that inflicted the actual injury or damage. For example, a worker was ripping planks on a 12 inch (30 cm) circular saw. To speed up production he removed the machine guard, thus exposing the blade. During the cutting process, he was distracted, and the blade cut his finger. In this case, the agency is the circular saw, and the blade would be classified as the agency part.

Agency Trends

A trend analysis can be made by listing the agency that causes the injury, disease, or damage. Trends can be used to establish which agency is responsible for the majority of losses as a result of contact.

Two Types of Agencies

There are two major classifications of agencies that are involved in the exchange of energy.

Occupational Hygiene Agencies: Occupational hygiene agencies are those items that are closest to and cause the illness or the occupational disease. They include:

- gas
- heat
- noise
- fumes
- radiation
- ergonomics
- lighting
- chemicals, etc.

General Agencies: These include the following:

- walkways
- machines

- ladders
- sharp edges
- power tools
- machinery
- equipment, etc.

OCCUPATIONAL DISEASE

An occupational disease is defined by McKinnon (1995) as:

> An occupational disease is a disease caused by environmental factors, the exposure to which is peculiar to a particular process, trade or occupation and to which an employee is not normally subjected or exposed to outside of or away from his normal place of employment.

(p. 27)

LEADING INJURY STATISTICS

In the National Safety Council's *Injury Facts, 2020,* the leading causes of injuries that required days off from work were:

- Exposure to harmful substances or environments
- Overexertion and bodily reaction
- Slips, trips, and falls.

According to Injury Facts 2020:
Exposure to harmful substances or environments includes:

- Exposure to electricity
- Exposure to radiation and noise
- Exposure to temperature extremes
- Exposure to air and water pressure change
- Exposure to other harmful substances
- Exposure to oxygen deficiency
- Exposure to traumatic or stressful event

Overexertion and bodily reaction include:

- Non-impact injuries
- Repetitive motion

Slips, trips, and falls include the following types of events (Figure 8.2):

- Slips and trips without falling
- Falling on the same level
- Falling to a lower level
- Jumping to a lower level (NSC Website, 2022)

Impact, Exposure or Energy Exchange	Losses	Costs
Driver struck against cab of tow motor	Injury to employee	$28,000
Tow motor struck against other carriages	Property damage to tow motor	$10,890
Tow motor and carriage fell to a lower level	Equipment damage to motor and carriages	$3,000

FIGURE 8.2 Showing an energy exchange and contact analysis after an accident investigation.

CONTACT CONTROL

Management has three opportunities to exercise the safety management function of control within the cause, effect, and control of accidental loss sequence (CECAL). The first opportunity is pre-contact control, which is setting up controls within the health and safety management system (SMS), identified by risk assessment, to manage the safety activities on a day-to-day basis. This reduces the root causes of accidents and in turn leads to a reduction in high-risk behaviors and high-risk conditions. This pre-contact effort will eliminate the accidental exposure, impact, or exchange of energy that causes the loss.

The second opportunity for control is during the contact stage. Control during the contact stage of the accident sequence can only be directed at minimizing or averting the amount and type of energy exchange. Contact control does not stop the sequence of events that leads up to the exchange of energy. It only deflects or transfers the amount of energy in another direction so that it does not cause harm to the body or structure.

As Bird and Germain (1992) describe contact control:

> Control measures that alter or absorb the energy can be taken to minimize the harm or damage at the time and point of contact. Personal protective equipment and protective barriers are common examples. A hard hat, for instance, does not prevent contact by a falling object, but it could absorb and/or deflect some of the energy and prevent or minimize injury.
>
> *(p. 26)*

Many SMS are focused entirely on the wearing of personal protective equipment, which will deflect an amount of energy but will not prevent the undesired event from happening. Nor will it stop the agency being in a position to transfer that energy to a person, piece of equipment, or the environment. Contact control is normally resorted to as an absolute last effort to minimize the degree of energy exchange.

EXAMPLE

Underground mining creates numerous hazards. A constant threat to underground miners is the possibility of head injuries caused by falling rocks and stones from the roof of the various tunnels in which they work. Although every precaution is taken to secure the roof, rocks and pebbles do fall and as a secondary measure, the wearing of hard hats is compulsory underground. This is a form of contact control.

Contact control should never be relied on as the first line of defense. As explained earlier, losses may be as a result of multiple contacts within the loss causation sequence. Thorough and effective accident investigation should identify all exchanges of energy that took place as well as all agencies and agency parts that were involved in causing the loss.

The third opportunity for management control is post-contact control which are all the actions and activities that take place after the loss.

EXPOSURE, IMPACT, AND CONTACT CONSEQUENCES

Exposure, impact, or contact with a substance or source of energy can cause numerous forms of loss. Most common are:

- injury to workers
- Illness or disease
- damage to property and equipment
- business interruption
- damage to the environment

Once an exposure, impact, or exchange of energy takes place, the end result and extent of loss is difficult to predict. As with the outcome of a high-risk behavior and/or high-risk condition, the outcome is largely fortuitous. By the time the contact takes place, the sequence of events (like the fall of the dominoes) is already progressing so fast that intervention at this stage is too late and some form of loss is imminent.

SUMMARY

This unplanned and unpredictable exchange of energy in the form of exposure, impact, or contact with a source of energy causes the accidental losses. There are numerous types of contacts and energy exchanges and the degree of, and type consequence cannot be predicted. Luck Factor 2 determines the outcome of this contact which could be injury, illness, property damage, business interruption or a combination. The outcome of this energy exchange could be minor, major, or catastrophic.

Part 2

Accidental Loss – The Effect

9 Luck Factor 2

INTRODUCTION

The failure to identify hazards, assess the risks, and institute the necessary controls causes accident root causes to exist. This in turn creates the high-risk behavior and high-risk condition which, as Luck Factor 1 would have it, results either in a near-miss incident (no loss) or an exposure, impact, or contact with a source of energy (loss).

The degree of exposure, impact, or contact is what causes the loss. The outcome of an exposure, impact, or exchange of energy is determined by fortuity or Luck Factor 2. The outcome may be injury to employees, damage to property, business interruption, or a combination (Figure 9.1).

TYPES OF ACCIDENTAL LOSS

Once the exposure, impact, or exchange of energy takes place, the outcome cannot be accurately predicted. Once there is an exposure, impact, or exchange of energy we have no control over the results.

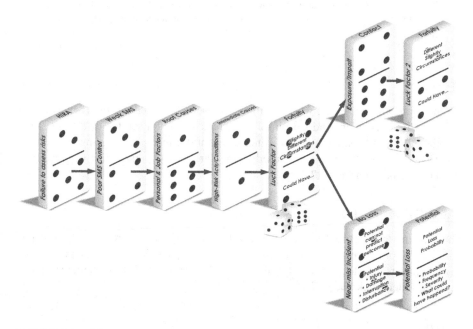

FIGURE 9.1 After an exposure, impact, or exchange of energy, Luck Factor 2 determines the outcomes.

DOI: 10.1201/9781003385943-11

The result of this exposure, impact, or exchange of energy, if above the threshold limit of the body or the substance, could be:

- disease or illness
- personal injury
- property damage
- business interruption
- or a combination of all of the above

There are numerous other sub-categories of loss such as environmental pollution, business disruption, and delays that are all some forms of accidental loss. The four main categories are discussed here.

TRADITION

Traditionally, most safety efforts were directed at one specific type of loss – injury to employees. Only with the introduction of loss control techniques were efforts focused on other consequences of an undesired exchange of energy. Organizations started to recognize and accept property damage accidents as accidental loss events even if there was no injury. Since occupational diseases are chronic and the symptoms often surface long after the exposure, work-related diseases did not get the attention that immediate injuries did.

Elsabie Smit and N.I. Morgan (1996) quote McKinnon:

> After the contact, luck again plays a role in determining the degree of severity of the contact. The outcome could be injury to people, damage to property or process interruption.
>
> *(p. 281)*

EXAMPLE

A question often raised is, why is it that some vehicle drivers write off their motor vehicles in an accident and get out of the wreck without a scratch or a bruise, and on the other hand somebody has a minor accident (for instance bumping into a lamp post), and minimal damage is caused to the car, yet the person dies because they broke their neck on impact?

ACCIDENT RATIOS

In referring to the Bird and Germain accident ratio discussed previously, the determining factor between the serious injury, the 30 property damage accidents and the events that resulted in minor injuries, is largely fortuitous. This is depicted by the domino titled Luck Factor 2.

SWIFT

SWIFT (Structured What if Technique) is a hazard identification method used to endeavor to predict the outcome of an undesired event. It is a prospective hazards analysis method that uses guide words and prompts to identify risks, with the aim of being quicker than other more intensive methods. It is a systematic system of hazard analysis that uses brainstorming techniques to consider the deviation, the hazard, and the cause, and endeavors to determine the consequence, if it should happen. This predictive method fits in well with risk assessment applications and reduces the uncertainty of outcomes of undesired events.

EXAMPLES

Take the example of a worker walking below an unguarded scaffold when a brick is accidentally bumped and falls to the ground. In the first instance, the brick falls to the ground and causes neither damage nor injury but minor process interruption, as the brick has to be picked up and returned to its original position.

In the second scenario, the same undesired event occurs and the brick falls, this time breaking and creating a loss in the form of damage to material.

In case three, the exact same undesired event takes place, the brick falls from the scaffold and this time hits the worker who happens to be passing by below. The exchange of energy causes minor injury.

In all the three instances given above there was an exchange of energy, yet the three outcomes, or losses, were totally different. This is as a result of Luck Factor 2.

It is extremely difficult, and in some cases impossible, to determine the outcome of an undesired exposure, impact, or exchange of energy. One factor that is inevitable is that the exposure, impact, or exchange of energy *will* result in some form of loss, the type, and the severity of which is largely fortuitous.

RESULTS NOT ACCURATELY DETERMINED

Once there is an inadvertent flow of energy and exchange, the results cannot be accurately determined. If an occasion occurs and no serious damage or consequence is realized, then the event itself is usually ignored as not being important. Predictive safety techniques such as SWIFT (Structured What if Techniques) and risk assessment help to project what could happen under slightly different circumstances. These circumstances cannot be predetermined except through intense risk analysis and hazard and operability (HAZOP) studies. If this is not done, and the event reoccurs, the outcomes are fortuitous.

SPACE SHUTTLE COLUMBIA

The National Aeronautics and Space Agency (NASA) space shuttle program is an example. On a number of shuttle launches, chunks of insulating foam broke away and some actually struck the craft. On one occasion this shedding foam damaged a heat tile of the Atlantis space shuttle during a launch.

After the launch of Atlantis, the NASA space shuttle program manager said an initial look at video and images of Atlantis' launch recorded five instances of foam shedding, but none were suspected of damaging the spacecraft.

The damage was insufficient to cause the shuttle to fail on reentry or landing.

On almost 80% of shuttle launches, foam strikes had been observed. Little action was taken to prevent this recurrence. On the final launch of the Columbia shuttle, a suitcase size chunk of foam hit the leading edge of the wing at the correct velocity and angle (slightly different circumstances) and caused sufficient damage to cause the craft to burn up on re-entry.

The CAIB report states that:

> Further photographic analysis conducted the day after launch revealed that the large foam piece was approximately 21 to 27 inches long and 12 to 18 inches wide (40x60 cm), tumbling at a minimum of 18 times per second, and moving at a relative velocity to the Shuttle Stack of 625 to 840 feet per second (416 to 573 miles per hour) (800 kph) at the time of impact.
>
> *CAIB Report*

Foam chunks passing the craft at high speed during launch creates a flow of energy but no loss. These are near-miss incidents. Foam chunks actually striking the craft and not doing damage are also near-miss incidents where the energy exchange was below the threshold level of the object. Foam striking the craft and damaging the wing structure is a property damage accident. In the Columbia instance we see a combination of all scenarios.

Many launches were unscathed despite foam chunks being shed. On the final launch however, the circumstances were right, and the foam chunk struck the leading edge of the wing, damaging the heat absorbing tiles which led to a burn up upon reentry into the earth's atmosphere.

HIRA OF VITAL IMPORTANCE

Hazard identification and assessing the risk of those hazards using various predictive techniques are what prevent the accident chain reaction occurring. It's too late once unplanned energy flows occur. The outcomes are largely fortuitous.

More Examples

The following four examples are real-life accident case studies. To emphasize the unpredictability of the outcome of accidents, the scenarios are given to show how Luck Factor 2 determines the result. Sometimes the consequence is an injury, property damage, business interruption, or a combination of all three.

Case Study 1: The steel cable came lose and swung down hitting a worker on the side of the head.
    This accident resulted in a minor injury.

Case Study 2: We were stripping steel scaffolding and there was a 440-volt cable on the ground. I placed a walk-board on the ground to kneel down and grabbed the scaffolding with my right hand. Just then the scaffold plank fell from above and narrowly missed me. The plank broke with the impact.

The result of this event was a heavy scaffold plank that fell to a different level and broke.

Case Study 3: The vehicle reversed into the bay and caught an electrical cable that was not secured to the side of the wall. The cable broke causing a power outage and business interruption.

The result of this exchange of energy was a damaged cable and a business interruption due to the power being disconnected.

Case Study 4: While cleaning contacts in a 480-volt main breaker cubicle, an electric arc occurred causing a first-degree thermal burn to the electrician's right wrist and a minor flash burn to his neck below his chin. The switch was damaged, and the milling machine was out of action for a few hours.

This accident resulted in injury, damage, and business interruption as a result of the flow of energy.

In all four accidents, there were different outcomes. It can be agreed that all scenarios had high potential for loss and that high-risk conditions or high-risk behaviors had resulted in some form of exposure, impact, or exchange of energy. The outcomes were fortuitous.

SUMMARY

Once an inadvertent and unplanned exchange of energy takes place, the outcome is unpredictable and could result in either injury or disease to persons, damage to property and equipment, or some form of business disruption. Pollution and other forms of loss may also occur as a result of this energy transfer.

The results of an accidental exposure, impact, or contact with a source of energy are unpredictable and the outcomes are normally as a result of chance or Luck Factor 2.

10 Injury, Illness, and Disease

The exposure, impact, or contact that leads to an exchange of energy can result in business interruption, damage to machinery, property, materials, and vehicles, or lead to personal loss in the form of injury or occupational disease.

Luck Factor 2 determines the outcome of the exposure, impact, or exchange of energy and this chapter will discuss the injury, illness, and disease outcome as a result of contact with a source of energy.

WORKPLACE INJURIES

The National Safety Council, *Injury Facts 2020* reports that there were 4,113 preventable injury-related occupational deaths during that year, and 4,000,000 preventable medically consulted injuries during the same period. This cost the US economy $643.7 billion (Figure 10.1).

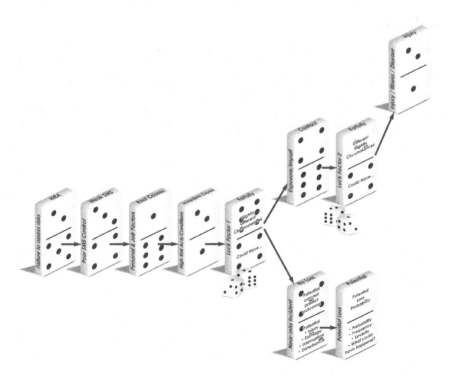

FIGURE 10.1 The exposure, impact, or exchange of energy results in a loss such as occupational disease, injury, or property damage.

DOI: 10.1201/9781003385943-12

MAIN FOCUS

The injury has traditionally been the main focus in loss-producing events. Most organizations still focus on the injury to employees as their main concern. Safety performance is normally measured by the number and degree of injuries that occur in a workplace.

Traditionally, an accidental injury was the only tangible proof that something had gone wrong with the process. It was also tangible proof of the consequences of either a high-risk work environment or a high-risk behavior. Injuries are more spectacular. They are instant, graphic results of the consequences of an accident.

COSTS

In numerous instances, the injury produced by an undesired event is not the most expensive consequence. Property damage, loss of production, business interruptions, etc., all caused by a similar sequence of events, normally cost more than injuries themselves.

POST-CONTACT

In the Cause, Effect, and Control of Accidental Loss (CECAL) sequence, the *injury* domino is in the *post*-contact or reactive stage. Although necessary, *post-contact* activities are far less efficient than *pre-contact* activities such as identifying hazards and assessing the risk, instituting the necessary controls, and ensuring by an ongoing measurement and evaluation system that the risk reduction controls are in place and are working.

INTERNATIONAL MISCONCEPTION

The misconception that exists is that a high number of injuries indicate "poor safety" and that an absence of injuries indicates "good safety." Research from authorities in this field will be quoted to show that injuries are used internationally to judge safety management success or failure. They are used for comparative purposes between organizations within the same state, between states, and even between organizations in different countries.

Having already analyzed the CECAL sequence up to this stage (injury, illness, and disease), it is obvious that the injury is a consequence that has already been determined by two luck factors. These factors determined the outcome of the high-risk action or high-risk condition as well as the outcome of the exposure, impact, or exchange of energy.

FORTUITOUS

The injury, therefore, is largely fortuitous and should not deserve the amount of attention that it gets in the safety management process. Admittedly, the injury is physical bodily harm to a friend, colleague, or fellow worker and one must be sympathetic to the pain and suffering that this person will undergo as a result of the injury. The emotion and feeling for the injured person should not cloud our vision and understanding that the injury was merely one of the events in a chain reaction that could have been prevented by proper health and safety management system (SMS) controls.

VICTIM

The injured person is the victim in the accident sequence. Being injured as a result of an accident does not necessarily mean that the injured person was the cause of the accident. Often the victim is automatically blamed for being injured. This means that, irrespective of the true and root causes of the accident, the victim bears the brunt of the accusation.

Cleverly worded accident investigation reports can point fingers at the victim irrespective of the circumstances of the accident. The fact that the victim may have committed a high-risk behavior does not necessarily justify blaming him or her for the accident.

Some of the classic excuses given on accident investigation reports are so simplistic and obviously blame finding that it is amazing management has accepted them.

EXAMPLES

A construction worker was working on the ground level where his task was to load bricks onto a hoist that then hoisted these up to the bricklayers on the next level. One of the bricklayers on the first level had not erected a toe-board on the scaffolding. During the course of the day, he accidentally kicked a brick, which slid off the scaffold and fell onto the worker below, severely injuring his arm.

On reading the accident investigation report, the builder foreman had given the cause of accident as, "employee took up a high-risk position and a brick fell onto his arm causing injury." Although it may sound ridiculous, this is an example of cop-outs or finger pointing directly at the victim.

Another example is taken from a deep-level gold mine where an employee was walking along a drift and a piece of the concrete lining the back wall fell on his head, causing him to fall and strike his knees on the ground. The accident report read as follows: "The employee failed to expect the unexpected." The injured employee is not necessarily the person who caused the accident. Thorough, factual, and accurate accident investigation will find the true causes of the circumstances that led to the injury.

OCCUPATIONAL DISEASES

Occupational diseases normally manifest over a longer period than occupational injuries. They are less spectacular and normally less apparent than an injury and have, regretfully, not received the importance that they should have in the safety management process. Perhaps a reason for this is that the words "accident" and "injury" have been made synonymous by safety researchers in the past. This has led the general public to believe that *safety, accidents,* and *injuries* are all together, as one, and occupational diseases and illnesses have been almost disregarded.

TANGIBLE RESULTS

Injuries are normally tangible consequences of an undesired event and happen quickly. Occupational illnesses and diseases are long-term and not so obvious. Noise-induced hearing loss, for example, normally occurs over a long exposure period and is not discernible by either the employee or his work colleagues. By the time the employee has suffered a hearing threshold shift, his hearing will have been permanently impaired to a degree that will affect his work and social life. Since there are no *immediate* consequences, this leads to a false sense of security, which often prevails in safety, that is, "Nothing will happen to me."

SYNONYMOUS

The words "accident" and "injury" have been incorrectly used in the past. *Accident* and *injury* have become synonymous and therefore the cause, effect, and control of accidental losses have also become confused. An injury is totally different from the exposure, impact, or contact with a source of energy, as are the other events leading to the contact. Heinrich et al. (1969) explain the confusion as follows:

> To the early safety practitioner, the terms "accident" and "traumatic injury" were almost synonymous. While occupational diseases, fire, and property damages were philosophically associated with industrial safety, actual accident prevention practices through the years have largely been devoid of these considerations and are quite injury oriented. Thus, the word "injury" has been most frequently used to mean bodily damage or harm through traumatic accident.
>
> *(p. 28)*

One of the biggest changes that the safety profession could make is to use the term *injury* in its correct context. An *injury* (harm) is not an *accident;* it may result from an accident (event).

INJURY DEFINED

Heinrich et al. (1969) explain *injury* as follows:

> Injury as used in this factor of the sequence, includes all personal physical harm, including both traumatic injury and diseases, as well as adverse mental, neurological, or systemic effects resulting from workplace exposure.
>
> *(p. 28)*

In *Injury Facts – Glossary* (2020), the National Safety Council defines *injury* as: "physical harm or damage to the body resulting from an exchange, usually acute, of mechanical, chemical, thermal, or other environmental energy that exceeds the body's tolerance" (NSC Website, 2020).

An *occupational injury* can also be defined as: "any injury such as a cut, fracture, sprain, amputation, etc., which results from a work accident or from exposure in the work environment."

An *occupational illness* can be defined as: "any abnormal condition or disorder other than one resulting from an occupational injury, caused by exposure to environmental factors associated with his or her employment."

The above definitions indicate that injury is some form of harm caused to a person's body. The harm could be acute or chronic and is normally caused by an accident. It is clearly a separate entity from the accident, which is the *event* that caused the harm.

INJURY STATISTICS

Most safety performance statistics concern injuries. Statistics based on injuries are also a traditional approach to justifying the need for better safety measures. It is pleasing to note that some statistics include damage to equipment, loss of production, and cost associated with environmental losses. These are also caused by accidents. Injury statistics are used as attention-getters. They quantify the number and extent of the injuries and diseases caused by workplace accidents and accidental exposures.

SOME INJURY FACTS

Injury Facts 2020 (National Safety Council) states that:

> There were 4,113 preventable work deaths in 2020 which is an increase of 5% over the previous year. There were also 4,000,000 medically consulted injuries which costs the economy $643.7 Billion.
>
> In 2020, the industry sector experiencing the largest number of preventable fatal injuries was construction, followed by transportation and warehousing. The industry sector experiencing the highest fatality rates per 100,000 workers was agriculture, forestry, fishing, and hunting, followed by transportation and warehousing.
>
> The leading cause of work-related injuries and illnesses involving days away from work in 2020 was exposure to harmful substances or environments. Exposure to harmful substances or environments was previously the 6th ranked cause. The next two most prevalent causes of injury and illnesses involving days away from work were overexertion and bodily reaction and slips trips and falls. These top three causes account for more than 75% of all nonfatal injuries and illnesses involving days away from work.
>
> *NSC website (2022)*

THE GOLD BARONS

In South African gold mines during the first 93 years of this century, about 69,000 miners died in accidents and more than a million were seriously injured. The 1986 Kinross mining disaster, when 177 workers were killed as a result of a polyurethane fire, was a notorious example of poor safety standards in the industry.

The commission into the accident conceded that mining is an inherently dangerous job, but it found that too often "profitability ranked higher than peoples' lives" – as evidenced by the asbestos scandal and the continued use of polyurethane in mines long after the dangers had become known.

World War II

Some attention-getting statistics concerning Americans killed in action during World War II are that about 373,000 persons were killed in the United States by accidents and about 408,000 were killed in war action.

ACCIDENT RATIOS

Although the accident ratios have been discussed previously, it is opportune to revisit them, as injuries form a major component of the ratio. Heinrich was the first to compile and publish the accident ratio. He estimated that for every serious injury there were 29 minor injuries. The Health and Safety Executive (HSE) (Great Britain) compiled its ratio in 1993 and found 11 minor injuries for every serious injury. Frank Bird and George Germain propose a ratio that indicates for every serious injury there were ten minor injuries. Figure 10.2 shows an actual accident ratio from a rod plant.

As mentioned previously, it is always the severe injury that gets management's attention. The minor injuries are normally referred to as first-aid cases, or injuries not requiring medical treatment, and often these injuries do not play a significant role in determining SMS effectiveness or non-effectiveness.

The accident ratios therefore indicate that approximately only 2% of all accidents result in injury. In the Bird ratio, 0.15% of all undesired events result in serious injury and only 1.6% in minor injury. This shows that injuries are, but a small percentage of total losses caused by accidents (Figure 10.3).

REPORTABLE INJURIES

Legislation normally defines what types of injury are reportable and what types need only be recorded in some form of register or log. These definitions differ among legal agencies, Workers Compensation, and insurance carriers. They differ from country

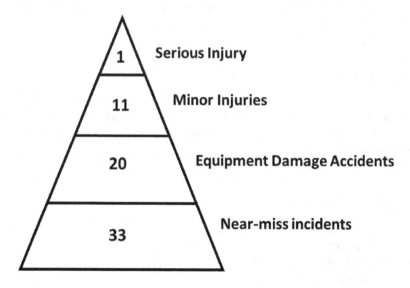

FIGURE 10.2 Accident ratio at a rod plant for ten months of Year 1.

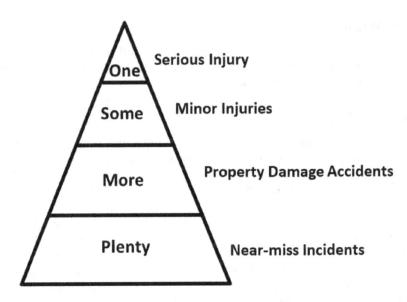

FIGURE 10.3 The updated accident ratio. (From McKinnon, Ron C. 2012. *Safety Management, Near Miss Identification, Recognition, and Investigation*. Model 2.4. Boca Raton, FL: Taylor & Francis. With permission.)

to country and are therefore not a reliable form of measurement at all. The very criteria that determine whether they are reportable are inconsistent.

NOT A GOOD INDICATION OF SAFETY

As these injuries and diseases are also the result of two luck factors, they are not a good indication of the work conditions or work practices at a workplace at all. Legal agencies would be far better off measuring the degree of controls that the organizations have put in place rather than basing the assumption of safety efforts on the degree of injury.

INJURY TYPES

There are a number of injury types that are usually grouped into 10 or 12 major categories. These give an indication of the type of injury being experienced for statistical analysis and tracking purposes. The term *injury* often includes occupational illnesses and diseases.

Some of the main injury types are as follows:

- Traumatic amputation
- Asphyxia
- Burn (heat or chemical)
- Contusion (bruise)
- Wound (laceration/abrasion)
- Skin irritation
- Dislocation

- Electrical shock
- Fracture (open/close)
- Frost bite
- Heat exhaustion/stroke
- Hernia
- Inflammation
- Sprain/strain
- Multiple injuries
- Puncture
- Foreign object

OTHER INJURY TYPES

Non-Impact Injuries

- Result from excessive physical effort directed at an outside source; common activities include lifting, pushing, turning, holding, carrying, or throwing.

REPETITIVE MOTION

- Microtasks resulting in stress or strain on some part of the body due to the repetitive nature of the task, typically without strenuous effort such as heavy lifting.

Occupational illnesses include diseases and disabilities caused by:

- Chemical agents
- Physical agents
- Biological agents and infectious or parasitical disease
- Respiratory diseases
- Skin disorders
- Musculoskeletal disorders
- Mental and behavioral disorders
- Occupational cancer

Although these classifications are by no means a complete description of the nature of injury (injury includes occupational illness and diseases) that could be experienced, they are useful in determining what type of injury or disease is most frequently caused by accidents. Some form of trend could be plotted using the nature of the injury or disease as a category. The nature of injury or disease would vary from industry to industry, and certain injury types are more frequently experienced in different industries.

MEASUREMENT OF SAFETY PERFORMANCE

Injuries are still used as the measurement of safety or the lack thereof. Organizations that experience fewer injuries than others are lured into a false sense of security by assuming that they are safer. Good safety controls are often *assumed* when an organization has low injury rates. Injuries are a poor indication of safety performance and are an even poorer indication of safety success. Injuries indicate a failure in the system and are more likely a measure of failure than of success.

The tradition has always been to rank a company's safety performance as well as its management's safety performance by the number of injuries experienced over a specific period. Internationally, this precedent has been set. It would prove to be exceedingly difficult to convince managers that the injuries they are using to measure their safety management controls are, in actual fact, merely indications of either good or poor luck. This is discussed in the Luck Factor 1 and Luck Factor 2 chapters.

Injury Statistics

Leading organizations and international companies quote injury statistics in their annual report to their boards of directors. They describe the disabling-injury incidence rate and the lost-time injury frequency rate as "the internationally accepted measure of safety performance." Although it is good that the boards of directors are receiving an annual safety report along with the financial report; it is time for more accurate, leading measurement of safety controls to be used.

LEGAL REPORTING REQUIREMENT

The reporting of certain injuries arising out of and during the course of normal employment is also a legal requirement in most countries. Most organizations require the reporting and recording of injuries. These are then tallied and used to compare one industry with the industrial average, or with other industries and companies belonging to the same organization.

V.L. Grose (1987) puts it in a nutshell when he says:

> Lots of thought and effort have been applied to managing risks over the past 30 years, but it has been only partially effective. Classical risk management's main drawback thus far has been its inability to participate as a full member of the top executive's team because it has not progressed much beyond the moral argument, "no one should suffer losses."
>
> To achieve their rightful status with top management alongside benefit managers, risk managers must merge the "flesh" or specialized activities of traditional risk management, with the "skeletons," or all-encompassing framework of the systems approach. The result can be called Systems Risk Management.
>
> *(p. 11)*

COMPETITIONS

Traditional safety competitions involve the comparison of injury rates between departments or companies or work sections. Prizes are then awarded to those divisions that have been injury-free or that have had fewer injuries. These incentives, which are still in practice today, have a major drawback especially if they are focused entirely on number of injuries. As Geller (1996) is quoted:

> Often incentives for fewer injuries, for instance, can reduce the reported numbers while not improving safety. Pressure to reduce outcomes without changing the process (or ongoing behaviors) often causes employees to cover up their injuries.
>
> *(p. 27)*

Mark A. Friend (1997), writing in *Professional Safety* continues the demise of traditional safety incentives:

> A typical safety incentive program is a prime example of how a firm creates an environment that fosters incidents. Under such a program, employees (or work teams) who do not report any injuries for a specified period of time are eligible for some reward. In this scenario, employees are encouraged to not report what is actually occurring.
>
> *(p. 36)*

He then continues and asks whether typical safety incentive programs do harm or good. He summarizes the answer by saying that typical programs reward employees for hiding facts – by encouraging them to not report injuries.

INJURY-FREE BONUSES AND INCENTIVES

When employees of an organization were interviewed, they admitted that there was a monthly safety bonus for being "injury-free." Paying safety bonuses in the form of cash incentives also leads to injuries being hidden and suppressed wherever possible.

Example: Questionnaire – Twenty-three subjects responded to a 15-question questionnaire. Two of the questions were directly pertinent to injury reporting. Question 1 asked, "If you work for a company that had worked 16,000,000 injury-free hours and you were injured in an accident, would you report it?"

Ten respondents said they would, 3 said 'no' they wouldn't and the other 10 said that, depending on the severity, they would either report it or not.

Question 10 asked, "If you were paid a monthly bonus of $250 cash to be free from injuries, would you report a personal injury?"

Three of the respondents replied "yes," 7 replied "no" they would not report it and the majority, 13 of the respondents, said that, depending on the severity, they would not report it.

In the chapter entitled "How we traditionally have analyzed" Dan Petersen (1996) also indicates the weakness of using the number of injuries as a measurement of safety:

> At what point do accidents become a valid measure to judge performance? Actuaries state that only when about 1,129 accidents have occurred would they begin to judge the unit that generated those 1,129 to be so believable that future rates could be based on those figures.
>
> *(p. 15)*

He states further that, "Using a frequency rate is a touch less ridiculous than this (a monthly fatality rate)" and asks the question, "So what do we use to measure the safety performance of line managers?" He answers the question by saying, "Almost anything but accident statistics." He continues, "How about an activity measure: did the manager do those defined activities that he or she agreed to? The activity measure would at least ensure some validity in the measurement."

(p. 16)

Writing in *Techniques of Safety Management*, Dan Petersen (1978) quoted further inaccuracies in injury reporting:

> The Department found significant problems and inaccuracies. For example, audit tests of 105 mines indicated that there could have been 118 more disabling injuries in addition to the 283 reported by the mines – an error rate of over 63%. It was also found that there could have been 108 more non-disabling injuries than the 139 reported by these companies – an error rate of 77%.
>
> *(p. 129)*

Dan Petersen and James Tye had obviously had discussions on under reporting and using injuries as a measurement of safety. Tye replied to a question about injury rates by saying, "since the under reporting of accidents in the UK runs at about 30% (and 80% in some U.S. industries), frequency rates, frankly, are irrelevant."

In Chapter 13 of *The Measurement of Safety Performance*, W. Tarrants (1980) discusses key issues and identified safety problem areas. Concerning the use of injuries as a measure he states that:

> Measures are needed that predict, not simply record, accident recurrences. Historical records of accidents mean nothing unless they can be used for prediction and control purposes.
>
> *(p. 235)*

He concludes by saying:

> A measurement problem exists in that low injury frequency rates tell nothing about the potential for catastrophe. For example, one plant experienced a $45 million accident loss, with its walls covered with safety awards received based on a low frequency rate. Better measurement techniques are needed to identify loss potential.
>
> *(p. 237)*

Most measurements using injuries are also based on the degree of severity of the injury and rely on the honesty of the reporting system. Most incentives drive injuries underground and management is lulled into a false sense of security thinking that its safety effort is having the desired effect.

Injury Not Reported

In one organization where the semi-annual production bonuses were dependent largely on the number of injuries experienced, an employee worked for six hours on a twisted ankle before finally collapsing in agony. When interviewed, he said he had tried to conceal the injury so that his team would not lose their portion of the bonus. His boot had to be cut off his swollen ankle before he could be treated. Peer pressure, team pressure, and fear of letting the side down and having them lose their bonus drove this man to walk on his injured ankle for hours before collapsing.

This is referred to as the safety fear factor – the fear of reporting a work injury because of the possible repercussions (Figure 10.4).

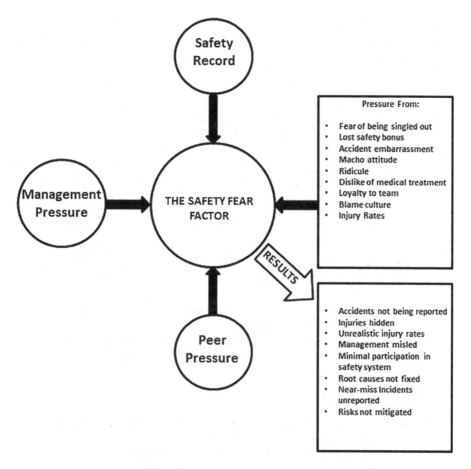

FIGURE 10.4 The safety fear factor. (From McKinnon, Ron C. 2014. *Changing the Workplace Safety Culture*. Figure 7.1. Boca Raton, FL: Taylor & Francis. With permission.)

Thomas A. Smith (1998), in his article, *What's wrong with safety incentives?* makes the statement that:

> Safety incentives/rewards create competition and fear within an organization. Although many believe that competition is positive, research dispels this myth. External competition will always exist, and it does drive business to perform better. Internal competition, however, is not positive. It creates winners at all costs. This is true in both production and safety. For example, an employee may not report an injury because he/she feels they are spoiling the department's record, which would mean no reward.
>
> *(p. 44)*

Injury Rates

Injury and illness rates tell the management of an organization where they have been. Risk assessment and reduction tells management where they are going.

The number of injuries is normally compiled and expressed as a percentage, or number of injuries per million workhours. The disabling-injury incidence rate is the percentage of employees suffering disabling injury per year. The lost-time injury frequency rate is the number of employees suffering lost-time injuries over a 1 million work-hour period.

More sophisticated injury statistics can be compiled by using the disabling-injury severity rate, which records the number of shifts lost per million or per 200,000 work-hours. By combining the severity and injury rate one can produce a lost-time injury index.

It has taken the safety profession many years to teach management that these statistics are the method to measure safety performance. Very few people who are exposed to these statistics actually know what they mean.

One of the shortcomings of using injuries as comparisons is that certain industrial workplaces and mines have more hazardous conditions, and it is therefore unfair to use the number of injuries as a comparative measurement.

Injury Analysis

The number, type, and body part injured are post-contact statistics that can be analyzed to plot certain trends. An example of the executive summary of an injury analysis report that covered 12 months of injury experience is as follows:

1. The first analysis was to identify employees who had been injured more than once during the last 12 months.
2. The next analysis was to confirm the department's DiiR (disabling injury incidence rate), and through this process, to determine the DiiR of individual departments and teams.
3. The day of the week and the number of injuries were analyzed and are shown in a pie chart format.
4. The time of day at which injuries occurred was analyzed and the results shown on a bar graph for the 24-hour cycle.
5. Injuries per department within the organization were also analyzed and displayed as total injuries per department on a bar graph.
6. Total injuries as per job classification were analyzed and shown on a bar graph. The exposure, impact, or contact types were analyzed in relation to the frequency and severity of injuries, namely disabling and non-disabling injuries.
7. The type of injury and the body part involved in these injuries is shown on two different graphs.
8. The total organization's injuries analyzed per first and second half of the year is shown.
9. Injury classification that constitutes the total injuries.

Report Summary

The report concluded with the following summary:

1. Total number of injuries reported was 81.
2. Fifty-three percent of the department's workforce reported an injury during this period.
3. Seventeen employees were responsible for reporting 45 injuries.

Analysis of Injuries: Part of Body

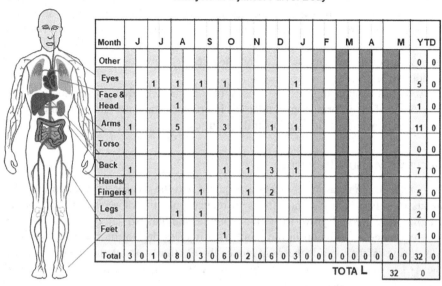

Month	J		J		A		S		O		N		D		J		F		M		A		M		YTD	
Other																									0	0
Eyes			1		1		1		1						1										5	0
Face & Head					1																				1	0
Arms	1				5				3				1		1										11	0
Torso																									0	0
Back	1								1		1		3		1										7	0
Hands/Fingers	1						1				1		2												5	0
Legs					1		1																		2	0
Feet									1																1	0
Total	3	0	1	0	8	0	3	0	6	0	2	0	6	0	3	0	0	0	0	0	0	0	0	0	32	0

TOTAL | 32 | 0

FIGURE 10.5 Number of injuries occurred to the body parts during the 12-month period.

4. Days of week injuries being reported showed nothing significant.
5. Most injuries occurred on day shift between 07:00 (7 am) and 12:00 (12 pm).
6. Laborers reported 52% of all injuries.
7. The leading energy exchange type was over exertion and the most severe was electric shock.
8. Parts of the body most frequently injured were back, hand and fingers, arms, and eyes.
9. Sixty-nine percent of injuries were recordable and 31% were first-aid injuries.

Figure 10.5 clearly shows that arms were injured more often than other parts of the body followed by back, hands, and fingers.

INJURIES ARE A SMALL PROPORTION

Injuries receive a lot of prominence in the safety management process. They are tangible consequences of the effects of accidental loss. Only a small percentage of accidents end up in injuries. Accident ratio studies have indicated that there are often more minor injuries experienced for every serious injury and that for the same serious injury there have been numerous property-damage accidents and a large number of near-miss incidents.

MEASURE OF SAFETY PERFORMANCE

Regretfully, injuries are still being used to measure safety successes and management's safety leadership. Due to the fact that they are largely fortuitous, as shown by the CECAL sequence, plus the fact that there is doubt as to the accuracy of the reporting of injuries, they are a poor indication of safety. Numerous safety-incentive schemes are a farce as they merely encourage non-reporting of injuries because of the benefits that can be derived in the form of prizes, awards, trophies, etc. Injury-free does not necessarily mean a high degree of safety control, nor does it mean accident-free.

Activities around injuries are mostly post-contact. Correct injury analysis and proper use of the information could assist in predicting and preventing further undesired events, which could lead to similar injuries.

ACCURATE MEASURE REQUIRED

A more accurate measure of safety is required and the measurement of certain elements of safety controls and efforts to reduce potential of accidental loss on an ongoing basis would prove to be far more substantial. No safety system or SMS can guarantee a reduction in the number of injuries due to the three luck factors. Measuring safety controls that help reduce the likelihood and that drive the risk into the *As Low as Is Reasonably Practical* (ALARP) area, are far more pro-active and accurate measurements of safety.

SUMMARY

Injuries to employees are one consequent of accidents. More property damage accidents occur than injury accidents, and there are always more minor injuries experienced for every serious injury recorded. The difference between a serious injury-producing accident and one that results in minor injury is often fortuitous. The same accident could have caused only property damage or resulted in a near-miss incident. Because of this, safety performance measured by injury rates is inaccurate and not a true reflection of the safety of the workplace.

11 Property and Equipment Damage

The exposure, impact, or exchange of energy caused by a high-risk behavior, high-risk condition, or combination thereof could result in injury, damage to property and equipment, or business interruption.

DAMAGE-CAUSING ACCIDENTS

There are numerous accidents that result in damage to property or equipment. This damage could extend to raw materials, vehicles, the environment, or finished products. Fire and explosions caused by accident are perhaps the most devastating types of property-damage events.

The damage is usually as a result of an exposure, impact, or contact and exchange of energy greater than the resistance of the item. The environment can be damaged as a result of both fire and pollution and extensive losses can occur even when no injuries take place. Most property-damage accidents also result in business interruption and some form of economic loss (Figure 11.1).

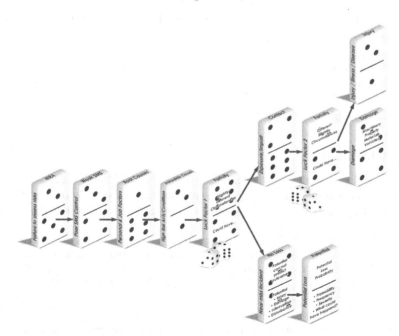

FIGURE 11.1 The one outcome of an exposure, impact, or contact with a source of energy could be property and equipment damage.

DOI: 10.1201/9781003385943-13

PROPERTY DAMAGE

Property-damage accidents are the most important in the accident ratio. They are warnings that a failure exists in the management system. Many property-damage accidents occur as a result of motorized vehicles colliding with buildings, the raw product, or the finished goods.

Property-damage accidents also have the potential to injure employees and therefore should not be ignored. All property-damage accidents should be thoroughly investigated, and a costing done to calculate the actual financial losses as a result of the accident.

Cost of repairs to equipment and vehicles should also be listed and tabled at the various safety committee meetings. These figures form a vital part of loss statistics.

COSTING

Some of the major types of property damage could include:

- buildings
- materials
- fixed equipment
- materials handling equipment
- motor vehicles
- tools

A simple classification for the level of property damage for investigation purposes could be:

1. Minor (less than $100)
2. Serious ($100 to $1,000)
3. Major ($1,000 to $10,000)
4. Catastrophic (over $10,000)

FLIXBOROUGH

The Flixborough disaster of 1974 has been termed one of Britain's greatest industrial accidents. In addition to the 29 people who died in the explosion, more than 100 were injured and 100 homes in the village nearby were destroyed or severely damaged.

> The blast took off slates and whole roofs in the village of Flixborough itself; all the windows were shattered; doors wrenched off and walls cracked. Chimney pots came tumbling down into the street and people were hurled about like rag dolls. Within moments, the peaceful scene resembled something out of the wartime blitz, said John Kennin, a witness.

Further damage was caused as a result of the explosion, as it totally destroyed Nypro's 18-million-pound (sterling) Flixborough plant, which was reduced to rubble as a result of the explosion.

SPACE SHUTTLE COLUMBIA

According to the Columbia Accident Investigation Board (CAIB), the space shuttle Columbia's demise was caused by a chunk of insulating foam, which dislodged from the left bipod ramp of the external fuel tank and struck the leading edge of the left wing, causing a breach of the thermal protection system. This in turn allowed super-heated air to enter the wing and melt the aluminum structure, which caused the wing to fail until aerodynamic forces caused the breakup of the orbiter.

The CAIB also found that foam had been shed on more than 80% of the 79 missions and that debris had caused damage on every space shuttle flight. There had been insulation shedding on most missions. After previous debris strikes, internal NASA investigations concluded it was an "accepted flight risk" and that it should not be treated as a flight safety issue.

A debris strike from the foam on the right SRB caused significant damage to the space shuttle Atlantis during the launch on December 2, 1988

The report also asks the question why the shuttle missions continued when they were aware of the foam being shed during ascents. The CAIB further stated that it would seem the longer the shuttle program allowed debris to continue striking the orbiters, the more opportunity existed to detect the serious threat it posed. One report indicates that engineers recognized that when foam hit the shuttle's wings upon lift-off, the shuttle was in possibly grave danger, but management failed to take the incident seriously enough (CAIB report).

Were damage accidents of foam striking the craft perhaps ignored? Shedding chunks of foam passing the craft at high speed with the potential to cause damage are examples of high-potential property-damage events and near-miss incidents.

ACCIDENT RATIO

In referring to accident ratios there are many more property-damage accidents for every accident that results in severe injury. This means that while some accidents result in injury, many more result in property or equipment damage.

LOSS CONTROL

The term "loss control" was first coined by Frank Bird in his book *Damage Control,* where it was stated that bridging the gap between traditional injury prevention programs and loss-control programs meant the recognition, investigation, and reduction of accidents that resulted in property and equipment damage. In shifting the focus from the tip of the iceberg i.e., the injuries, Bird and Germain (1992) explain the concept:

> Third, if the event results in property damage or process loss alone, and no injury, it is still an accident. Often, of course, accidents result in harm to people, property, and process. However, there are many more property-damage accidents than injury accidents. Not only is property damage expensive, but also damaged tools, machinery and equipment often lead to further accidents.

(p. 18)

They continued their motivation of the importance of the property-damage accident by explaining that investigation of property-damage accidents could lead to the root causes of the event that, under different circumstances, might have led to injury of personnel.

Frank Bird pioneered in the expansion of industrial safety from an injury-oriented concept to a discipline encompassing all accidents by his extensive studies and writing on the identification, costs, and control of all property-damage accidents during the 1950s and early 1960s. The book, *Damage Control*, co-authored by him and published by the American Management Association in 1966 was one of his many publications on this subject.

Bird and Germain conducted studies in 1954 and they bridged the gap between accident and total accident prevention. All too often companies failed to recognize the extent and frequency of industrial accidents. Usually only injury accidents are reported, investigated, and analyzed, and frequently others are ignored as near misses despite the fact that they indicate potential hazards and cost thousands of dollars in property damage alone.

The success that Frank Bird and George Germain (1992) had in introducing the concept of loss control is explained in *Practical Loss Control Leadership*:

> During this period *damage control* provided a logical bridge from injury-oriented safety programs to accident-oriented programs. More and more people recognized not only that accidental damage is extremely expensive, but also that damage accidents have significant potential for injuring and killing people. *Damage Control* is the first book published on a totally new approach to plant safety that places the emphasis on *all* accidents – not just those resulting in injuries. Today damage control is recognized as a vital part of safety/loss control by leading organizations around the world.
>
> *(p. 8)*

Bird and Germain (1992) further explain why property-damage accidents were seldom considered:

> As we considered a ratio, we observed that 30 property damage accidents were reported for each serious or disabling injury. Property damage accidents cost billions of dollars annually and yet they are frequently misnamed and referred to as "near-accidents." Ironically, this line of thinking recognizes the fact that each property damage situation could probably have resulted in personal injury. This term is a holdover from earlier training and misconceptions that led supervisors to relate the term "accident" only to injuries.
>
> *(p. 21)*

COSTS

Injury Facts 2020 edition, published by the National Safety Council, estimates that motor vehicle damage for the year 2020 was in excess of US$473 billion.

In the US alone, local fire departments responded to 1,338,500 fires in 2020. These fires caused 3,500 civilian deaths, 15,200 civilian injuries, and $21.9 billion in property damage. The same report quotes the cost of all injuries as $1,158.4 billion. Work-related injuries cost $163 billion during the same year.

MONTH	PROPERTY DAMAGE COST	MONTH	PROPERTY DAMAGE COST
January	$6,930	July	$8,709
February	$47,114	August	$19,124
March	$32,795	September	$3,006
April	$5,968	October	$14,775
May	$108,656	November	$8,580
June	$21,587	December	$250
		Total for the year	$277,495

FIGURE 11.2 Actual costing of equipment and property-damage accidents during one year.

Even in medium-sized organizations, property damage as a result of accidents should be tracked and costed to establish a full cost of these accidents. An example of costs incurred by property-damage accidents at a manufacturing plant is given in Figure 11.2.

These figures are taken from actual property damages reported by a company and give some indication of the extent of equipment and property damage that could be incurred by accidents.

IMPORTANCE OF EQUIPMENT AND PROPERTY-DAMAGE ACCIDENTS

As stated, property-damage accidents are caused by exactly the same sequence as injury-producing accidents. As they also have potential to cause injury, they should receive the same attention and investigation as injury-producing events. To determine the total cost of risk within an organization, the costing of all losses as a result of undesired events should be done monthly, and on a progressive basis.

This information should be circulated to management and departmental management and should be discussed at safety committee meetings. Although an organization may have relatively few injuries, their property-damage accidents may be costing a great deal (Figure 11.3).

The damage listed in the above investigation model was to the tow motor, the carriages, and the electrical system. Various items can be damaged during one accident and thorough investigation techniques should identify all exchanges of energy and all subsequent damages.

EXAMPLES

Some examples of property-damage accidents reported by one organization are given:

- The employee was passing by when he saw another employee in the lunchroom dozing off. When he hit the window to get his attention the window broke.

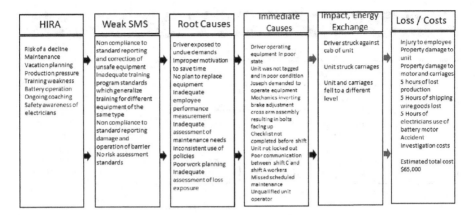

HIRA	Weak SMS	Root Causes	Immediate Causes	Impact, Energy Exchange	Loss / Costs
Risk of a decline Maintenance Vacation planning Production pressure Training weakness Battery operation Ongoing coaching Safety awareness of electricians	Non compliance to standard reporting and correction of unsafe equipment Inadequate training program standards which generalize training for different equipment of the same type Non compliance to standard reporting damage and operation of barrier No risk assessment standards	Driver exposed to undue demands Improper motivation to save time No plan to replace equipment Inadequate employee performance measurement Inadequate assessment of maintenance needs Inconsistent use of policies Poor work planning Inadequate assessment of loss exposure	Driver operating equipment in poor state Unit was not tagged and in poor condition Joseph demanded to operate equipment Mechanics inverting brake adjustment cross arm assembly resulting in bolts facing up Checklist not completed before shift Unit not locked out Poor communication between shift C and shift A workers Missed scheduled maintenance Unqualified unit operator	Driver struck against cab of unit Unit struck carriages Unit and carriages fell to a different level	Injury to employee Property damage to unit Property damage to motor and carriages 5 hours of lost production 5 Hours of shipping wire goods lost 5 Hours of electricians use of battery motor Accident Investigation costs Estimated total cost $65,000

FIGURE 11.3   A basic CECAL model used to investigate an injury and property-damage accident.

- I stopped the truck and got off to direct the mechanic on the forklift and I thought I had put the gearshift into park. The truck rolled and ran into a flatbed trailer.
- The riggers were checking the cables on the auxiliary lift and the block was lowered to the ground. Excessive cable was unwound off the drum and this came into contact with an energized electric welding lead lying on the ground. This burnt a portion of the auxiliary hoist's cable.
- A crane knocked the handrail off the deck and pulled out the electrical cable for the controls of the mixing machine.
- A forklift truck was stuck. A second forklift driver picked up the back of the stalled forklift truck and got his forks too far under it and put a hole in the oil pan.
- The crane driver was going to the south end of the building to get a drink of water. He thought that the crane was in the neutral position and when he approached the end of the building he looked down and noticed the control was in the drive position. He moved the control back but there was not enough time. The crane hit the wall.

These property-damage accidents clearly indicate the potential for injury and also the costs of repairing the accidental damage as well as the accompanying disruption.

NON-REPORTING

The same reasons for the non-reporting of injuries exist for property-damage accidents. Top of the list is a fear of disciplinary measures as property-damage accidents are expensive and result in disruption. A good SMS should encourage all property damages to be reported. A costing of the accident should be done and any damage in excess of $1,000 deserves a thorough investigation and attempt to prevent a recurrence of a similar type of accident. The events leading up to a property-damage accident should be investigated as if it were an injury-causing accident.

CONCLUSION

Accidental losses could include losses to people, equipment, material, or the environment. Many equipment and property-damage accidents have potential to cause injury to workers under different circumstances. Since these non-injury accidents have potential for injury and create a loss, they should be investigated, costed, and treated with the same amount of diligence as injury-producing accidents.

12 Business Interruption

An *accident* has already been defined as: "an undesired event that results in harm to people, damage to property or business interruption." Business interruptions are perhaps the most intangible consequence of accidents and are seldom treated as true accidents (Figure 12.1).

A *business interruption* is defined as: "an undesired event that does not result in injury or damage, but which produces an undesirable change to the normal flow of process or manufacturing, thus interrupting the business."

LUCK FACTOR 2

As already explained, Luck Factor 2 determines which of the three categories of loss will be produced by an accidental exposure, impact, or contact with a source of energy. They are:

- injury or disease
- property and equipment damage

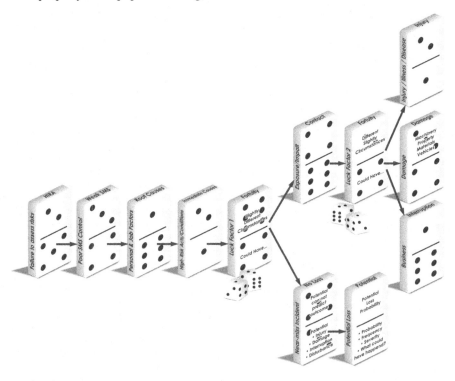

FIGURE 12.1 The domino depicting the loss due to business interruption.

 DOI: 10.1201/9781003385943-14

- business interruption
- a combination of either two or all three

INJURY ACCIDENTS GET MORE ATTENTION

Injuries, occupational diseases, and property damage tend to receive more attention than business interruptions or loss of process. Loss of process or business interruption is normally accepted as part of doing business, even though the events leading up to the business interruption are identical to those leading up to injury or property damage. Traditionally, there has been little concern shown for, or recognition of the fact that business interruption and loss of process could result from an accident. Every injury or property-damage accident interrupts the business.

ALL ACCIDENTS CAUSE LOSS

All accidents, whether they result in injury, property damage, or business interruption, cause some form of loss. An interruption of normal day-to-day business can also be caused by an accident. Some accidents, however, do not cause injury or result in any damage whatsoever, but stop or retard the intended outputs of the organization and thus interrupt or hinder the process.

This interruption could be in the form of:

- lost time
- lost product
- loss of process
- inferior quality
- rework

These business interruptions as a result of accidents are seldom, if ever, reported and investigated and consequently very seldom there is an effort made to reduce their causes. Recently there has been a shift in focus from the injury to property damage, as a more frequent, and just as important, consequence of an accident. A further thrust is needed in the future to broaden the focus to include the loss of product and business interruptions caused by the same series of events that caused other losses.

A CAN OF WORMS

As with root cause analysis, investigation, and costing of business interruptions due to unplanned events could be like opening a can of worms, as these interruptions clearly indicate a breakdown of management control and supervision.

LOSS CONTROL

Control of accidental loss is a part of every manager's job, and the reasons for this are that managers are responsible for the health and safety of others. Safety management provides an operational strategy to improve overall management, and managing safety provides significant opportunities for managing costs.

Many management specialists agree that about 15% of a company's problems can be controlled by employees, while 85% can be controlled only by management. This means, in other words, most safety problems are management problems.

OPERATIONAL PROBLEMS

Since the management system is responsible for budgeted and predicted output of an organization, any interruption would reflect directly on a deviation from intended management activity.

Dan Petersen (1978) also linked accidents to weaknesses in the management system:

> The root cause of accidents (weaknesses in the management system) is also the cause of other operational problems. Though this fact is not immediately obvious, the more we consider it, the more obvious it seems to become. Consider, for instance, how often our safety problems stem from lack of training – and how often our quality problems also stem from the same lack of training. Or consider how poor selection of employees creates safety problems and other management problems. The fundamental root causes of accidents are also fundamental root causes of many other management and operational problems.
>
> *(p. 42)*

Many agree that the methods of most value in accident prevention are analogous with the methods for the control of quality, cost, and quantity of production.

QUALITY

The sequence of events that normally affects the quality of the process as well as the quality of the end product is remarkably similar to the sequence that results in injury to people. Failing to identify the hazards and assess the risk of undesired events occurring in the planning, production, and marketing of products could also result in inferior quality and expensive product recalls required by law.

WORKER MORALE

Simonds and Grimaldi (1963) discuss the peace of mind of the worker being affected as a result of accidents, which could lead to slowdown in the work throughput:

> In a no-injury accident the factors of unpleasantness and possible temporary physical impairment would not be so cogent, but even then, according to the definition, there must be an element of danger. Therefore, even no-injury accidents are not likely to be conducive to the peace of mind of the workers.
>
> *(pp. 86–87)*

This means that a slowing down or extra caution exercise by workers as a result of an accident could start to hinder the normal production. They continue discussing the costs related to accidents as follows:

If production lost because of an accident is made up by working overtime, the accident should be charged with the difference between the cost of doing the work in overtime and the cost that would have been incurred if it had been done in regular hours.

(p. 88)

Here they are referring to interruptions or delays and time lost as a result of an interruption caused by an accident. An interruption could also have a chain reaction and affect other workstations, especially in production-line-type industries.

This accident costing would include workers who are forced into temporary idleness by the accident but are still being paid their normal wage. It would also include their cost of supervision having to investigate and rectify the business interruption as one of the hidden costs of business interruptions.

Other costs connected with business interruptions could include:

- contracts canceled
- orders lost if the accident causes a long delay
- reduction in total sales
- loss of bonuses
- cost of hiring new employees

INCREASED OR DECREASED PRODUCTIVITY

In a three-year study on the impact of a good health and safety management system (SMS) on productivity, a warehousing facility quotes the following statistics:

As a result of the introduction of a safety management system, the company experienced:

- a total accident frequency reduction of 42%
- the disabling injury frequency reduced by 81%
- labor turnover rate was reduced by 34%
- plant availability increased by 16%
- production output increased by 39%
- loads per employee increased by 17%
- machinery availability increased by 2%
- degree of safety on the total organization was up 24%

A substantial drop in accident injury frequencies with corresponding savings in costs, time, and production shifts, as well as improved employee safety culture, all attributed to the implementation of the SMS. It proved positively that safety and productivity go hand in hand.

BUSINESS INTERRUPTION

A *business interruption* has already been defined as: "an undesired event that does not result in injury nor damage, but which produces an undesirable change to the normal flow of process or manufacturing, thus interrupting the business."

If one couples this definition with the definition of *accidental loss* which is: "an avoidable waste of any resource," it becomes clearer that focusing on accidents that interrupt the business could identify weaknesses in the management system and rectification measures could in turn improve both quality, output, and production.

Example

In an organization that relies on tonnage produced, a study was undertaken, and it was found that ~10% of the process was lost as a result of accidents.

An analysis of the loss of product was undertaken and it was discovered that 24% of the product was being lost due to spills, 10% lost to mechanical failures, and 7% as the result of motor failure. A further 30% was lost as a result of miscellaneous causes. Further investigation showed that 90% of the causes for the lost process were accidental and could have been prevented.

Figure 12.2 shows an analysis which was taken over a six-month period. The results show the percentage of process lost, as well as a breakdown of the causes of the production being lost.

Assessing the risk and instituting control measures can prevent the majority of all undesired events. This means those business interruptions, loss of process, and inferior quality can be reduced and controlled in the same way as undesired events that result in injury.

REASONS FOR LOST TONNAGE

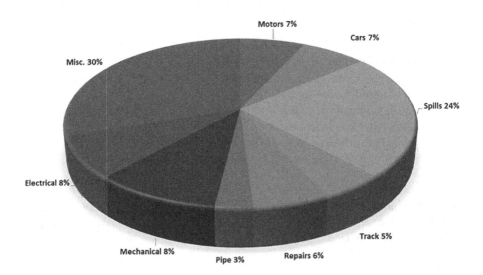

FIGURE 12.2 A break-down of accidental process lost per month.

SUMMARY

An accident can result in one or a combination of three major consequences.

1. Occupational injury or illness
2. Property or equipment damage
3. Business interruption in the form of loss of process, low productivity, or inferior quality

Business interruption as a result of an accident receives less attention than spectacular property damage and injury to employees yet occurs more frequently than the latter. In relating to the accident ratio conclusion, if there are more property-damage accidents for every serious injury, there must be at least hundreds of events that disrupt the normal output of an organization.

Business interruptions could mean workers are idle, machines are not fully utilized, time is wasted, the number of units produced is reduced, or the intended production flow is hampered.

Business interruption accidents should be costed, and their causes identified and rectified. As with property-damage accidents, business interruption accidents do have potential to cause damage and/or injury. Production management should include the avoidance of business interruption accidents in the same way as accidents that cause injuries to employees.

13 Luck Factor 3

Heinrich et al. (1969) compiled ten axioms of industrial safety, the most pertinent one to this chapter being axiom number 4, which states *"the severity* of an injury is largely fortuitous – *the occurrence* of the accident that results in injury is largely preventable" (p. 21).

This fourth axiom is perhaps the most significant statement in the safety management profession. What Heinrich wrote is that the degree of injury depends on luck, but that the accident can be prevented. What he further indicates by this axiom is that while the accident can be prevented, the severity of the resultant injury is something over which we have little or no control (Figure 13.1).

FOCUS ON INJURY

In examining the Cause, Effect, and Control of Accidental Loss sequence (CECAL), once an injury occurs as a result of an exposure, impact, or exchange of energy, the degree of injury is largely dependent on fortuity. Most safety activities are focused on the severity of an injury or illness. Consequently, the focus is on an end result, which is determined by fortune, chance, or luck.

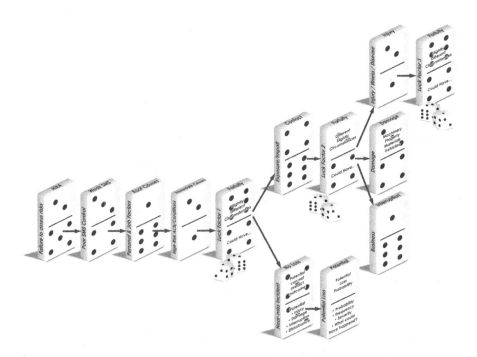

FIGURE 13.1   The domino depicting Luck Factor 3.

   DOI: 10.1201/9781003385943-15

In the fourth edition of *Industrial Accident Prevention*, printed in 1959, H.W. Heinrich and E.R. Grannis further explain Luck Factor 3 by stating:

> There are certain types of accidents, of course, where the probability of serious injury may vary in accordance with circumstances.
>
> *(p. 28)*

In other words, the circumstances that determine the probability of a serious injury are really those of either good or bad fortune.

They continue their discussion as follows:

> An injury is merely the result of an accident. The accident is controllable. The severity or cost of an injury that results when an accident occurs is difficult to control. It depends upon many uncertain and largely unregulated factors – such as the physical or mental condition of the injured person, the weight, size, shape, or material of the object causing the injury, the portion of the body injured, etc. Therefore, attention should be directed to accidents as properly defined, rather than to the injuries that they cause.
>
> *(pp. 28–29)*

PATTERN

All loss-causation events follow the CECAL pattern, but their progress through the loss-causation model is channeled either by Luck Factor 1, Luck Factor 2, or Luck Factor 3. The difference between a fatality, permanent disabling injury, temporary disabling injury, lost-time injury, and a first-aid case is largely a matter of luck.

In examining hundreds of cases where the degree of injury could have been far greater than it was, it is difficult to explain the resultant injury other than by conceding it was luck, as depicted by Luck Factor 3.

CASE STUDY

The following example shows how the degree of injury is fortuitous.

Electrical Contact Accident: Because the electrician assumed the contacts were de-energized, he did not check with the meter to verify the situation, and he did not take hot-work precautions. Using a wire cup-wheel in an electric drill, he was cleaning one contact when he apparently bridged between two of the three contacts initiating an arc that, once started, bridged all three phases. A witness stated that the arc flash enveloped the electrician. The arc caused burns to his wrist and neck. It also burned holes in his long-sleeved work shirt, scorching it all the way up both arms to the back of his neck; pitted his safety glasses; singed facial hair and hair on the back of his neck. It scorched papers protruding from his shirt pocket. The arc tripped a breaker, cutting power to the plant. The electrician was treated at the clinic, released back to his normal duties, and completed his scheduled work shift.

Because the electrician was not seriously injured, this was not recorded as a reportable injury, or a lost-time injury and the safety "record" was not affected at all. Reading the description of the accident it appears obvious that to all intents and purposes the electrician should have been electrocuted yet suffered only a minor injury.

Electrical experts say that any electrical current between 16 and 1,500 mA can cause various effects such as severe pain, paralysis, and ventricular fibrillation. They give a case history:

> A young girl, who suffered from a heart ailment, picked up a badly joined flexible lead of a bedside lamp with one hand. She was standing on a carpet and the burns on her hand indicated contact between the line and neutral at her hand only. Although the power was isolated within a very short time, the girl died.

The two accident scenarios clearly indicate that the degree of injury is difficult to predict and is therefore determined by other circumstances beyond our control.

CONTACT WITH ELECTRICAL CURRENT

While analyzing a minor injury log, it was found that there were four situations where the accident type was "contact with a source of electricity." Of these four different accidents, none resulted in serious injury, and no one was electrocuted or severely burnt.

Accident 1: "The pipe fell, slipped forward and cut the 480-V electrical cable, shocking some men."

Accident 2: "I was climbing the ladder near the crane gantry, and I felt an electrical shock."

Accident 3: "There was a 220-V cable on the walkway. I placed some walk boards down to kneel on them and grabbed the steel with my right hand while my left hand brushed the cable. I got shocked."

Accident 4: "I was changing the position of the stepladder, which hit an overhead electrical cable and shocked me."

Having contact with medium to high voltage and not experiencing a fatality is indeed luck. Ventricular fibrillation can occur from 15 mA upwards. The young girl who died as a result of a relatively insignificant contact with electricity was less fortunate.

Any measurement of safety performance based on degree of injury is based on the degree of *luck*. Admittedly, the number of serious injuries and fatalities is important, as is the number of first aid and dressing cases. Forward thinking specialists say that organizations should quit looking at injury-based measurements to assess system effectiveness.

WRONG EMPHASIS

So much emphasis is placed on *lost time, disabling* injuries, and *reportable* injuries that this degree of harm to the body has become the focus of most SMS, safety practitioners, unions, and regulatory bodies. If one accepts Heinrich and the other safety pioneers' statement that the degree of injury is fortuitous, then using the degree of injury as a measurement tool is really measuring an organization's *luck factor*. Even boards of directors, when quoting lost-time or disabling-injury rates in annual reports, regard these measures (of safety) as being "internationally accepted measurements of safety performance." They are measurements based on the degree of consequence after three factors of luck that have determined the outcome.

THE LUCK FACTORS IN THE ACCIDENT SEQUENCE

From the high-risk behavior or high-risk condition, Luck Factor 1 determines whether there is going to be an exposure, impact, or exchange of damaging energy. Once there is an exchange of damaging energy, Luck Factor 2 determines whether that energy will cause a business interruption, property damage, or personal injury/illness, or a combination of two or all three. Once that energy is directed toward a human body, Luck Factor 3 (or circumstances beyond our control and our prediction) determines how much energy is going to be exchanged with what part of the body, under what circumstance. The result could be anything from a minor injury to a fatal injury.

MEASURES OF FAILURE

Many professionals advise not to use the degree of consequence as a measure. James Tye openly stated that frequency rates were measurements of failure rather than success.

Some professionals ask that if you go from an occupational injury rate of 8 to 2 in one year, does this mean that you are so much better? If so, what is it that the company did to get better? Organizations usually say: "I have no idea. We were just lucky, I guess."

W. Tarrants (1980) also said that:

> To the extent that it can be argued that all injury-producing events are the consequence of the same causal structure, and that the degree of injury is merely a matter of chance, there are clear advantages in a measure that includes all events in the sample.
>
> *(p. 180)*

He continues by saying:

> At any future point in time, accidents that have a potential for loss may produce a loss. Which exposure results in loss and the degree of severity of future losses are influenced by chance factors.
>
> *(p. 15)*

What Tarrants is referring to in this statement is the three luck factors. These determine whether there is going to be a loss or a near-miss incident. If there is an exposure, impact, or contact with a source of energy, luck determines whether there is going to be injury or property damage and disruption, or if there is going to be injury, how severe it will be.

In emphasizing how Luck Factor 3 clouds management's vision as to the effectiveness of the SMS, Tarrants (1980) explains:

> One big problem is the accident (injury) that escapes detection because it appears below the fine dividing line established by the ANSI disabling injury or the BLS-OSHA recordable injury or illness definitions.

He continues by explaining why injury rates are not a good form of safety measurement:

> None of the measures of safety performance commonly in use is acceptable as a valid measure of identifying our internal accident problems. It is important that one is not lulled into believing that present measures are really descriptive of the actual level of safety effectiveness within an organization.
>
> *(p. 40)*

Continuing the justification of not using categories of injuries as a measurement of safety, Tarrants (1980) explains:

> A measurement problem exists in that low injury frequency rates tell nothing about the potential for catastrophe. For example, one plant experienced a $5 million accident loss, with its walls covered with safety awards received based on a low frequency rate. Better measurement techniques are needed to identify loss potential.
>
> *(p. 40)*

He continues his argument:

> Rates are too crude. They fluctuate too much, they are too unstable, and changes in the rate do not necessarily reflect changes in the level of safety within the organization measured. The rate can fluctuate drastically on a fortuitous basis. Likewise, measures of severity can fluctuate widely by chance. For example, a person can fall and bruise an elbow. One can also fall from the same height and fracture the skull and die. The cause can be the same, but the effects and ultimate results are substantially different.
>
> *(p. 239)*

Here Tarrants is once again emphasizing that the injury and severity are fickle measures of performance because of the chance outcomes as depicted by Luck Factor 3.

In a recent study undertaken by the Accident Prevention Advisory Unit of the Health and Safety Executive, of the United Kingdom, the extensive study was summarized as follows:

> Any simple measurement of performance in terms of accident (injury) frequency rates or accident/incident rate is not seen as a reliable guide to the safety performance of an undertaking. The report finds there is no clear correlation between such measurements and the work conditions, in injury potential, or the severity of injuries that have occurred. A need exists for more accurate measurements so that a better assessment can be made of efforts to control foreseeable losses.

SAFETY PARADIGM

The case against using the degree of severity of an injury as a safety performance measure is certainly strengthening. Safety pioneers acknowledge that the severity of an injury is fortuitous. They agree that a large factor of luck exists in determining the resultant injury. Yet safety practitioners, regulatory agents, and most management gauge their safety performance on the number of injuries that lose a shift or longer.

On achieving 1 million injury-free workhours, massive safety celebrations are held at factories, plants, and mines all around the world. Yet it has never been scientifically proven that the introduction of a safety-management control system has a direct bearing on the number or severity of the injuries. This was also stated by the Health and Safety Executive's intensive report. What is certain is that assessing the risk and instituting controls reduces the probability of high-risk behaviors and high-risk conditions occurring but because of the three luck factors, injuries *will* occur and, depending on luck, they may be disabling, minor, or even fatal.

Dan Petersen (1978) strengthens the argument by saying:

> Perhaps our biggest doubt comes when we ask ourselves whether a death count or death rate really reflects our national effort and performance in safety. We seem to believe that fatalities are caused largely by "luck," and thus we tend to raise our eyebrows at such a measure.
>
> *(p. 128)*

CHALLENGE

The challenge to the safety profession, and to management, is to stop regarding the count of a certain level of injury as a measure of good or poor safety performance. The degree of control, the activities, and ongoing processes and procedures within the SMS are a far better measurement of safety than the number of lost-time, reportable or disabling injuries, which are determined by luck.

Dan Petersen continues to explain Luck Factor 3 as follows:

> It measures a level of failure somewhat less than a fatality (an injury serious enough to result in a specific amount of time loss from work), but the fact remains that the supervisor of 10 workers can do absolutely nothing for a year and attain a zero-frequency rate with only a small bit of luck. By rewarding such a supervisor, we are actually reinforcing non-performance in safety.
>
> *(p. 66)*

QUESTIONNAIRE

In a written survey involving 23 employees, question 11 reads as follows: Do you believe that "luck" has a part in the outcome of an accident and the severity of an injury. Of the 23 replies all replied "yes" to both parts of the question.

CASE STUDY

In this particular fatal accident, the degree of severity was worsened by luck. The following factors were listed that could have worsened the degree of injury.

1. The victim was only released from the cab of the damaged vehicle 47 minutes later. The length of time to free the victim may have played a role in allowing the severity of injuries to increase.

2. The light was fading, which may have hampered the rescue operations.
3. Traveling time to the closest medical center was two hours. If a medical rescue helicopter had been summoned as soon as the accident happened, it could have been on site at the time of the victim's release from the cab.
4. If the vehicle had fallen on its other side, the degree of injury might have been less.

MIDDLEBULT DISASTER

The Middelbult Coal Mine disaster in 1987 is a prime example of this. Having been rated a five-star mine by NOSA (National Occupational Safety Association, South Africa) for three years preceding the accident, it had also maintained a disabling-injury incidence rate of less than one (<1% of the workforce being injured per year) for three years concurrently, when, in one underground explosion, 53 miners were killed.

Using the degree of injury as a measurement of safety performance or safety failure is completely misleading. The difference between minor injuries, disabling injuries, lost time, and fatal injuries is largely a matter of fortuity.

CONCLUSION

The type of injury that results from an accident is largely dependent on factors that can neither be predicted nor controlled, and which numerous safety authors refer to as fortune or chance.

H.W. Heinrich's (1959) one-line statement in *Industrial Accident Prevention*, "the severity of an injury is fortuitous," seems to have been grossly overlooked by the safety profession. They have ignored what he has said and based all efforts, recognitions, punitive measures, successes, and failures on the degree of injury. Numerous studies have also concluded that there is little, if any, correlation between good safety controls, a good SMS, and the number and severity of injuries.

14 Severity of an Injury

The severity of an injury is largely dependent on Luck Factor 3 and many safety pioneers and previous safety management research confirms this fact. The degree of injury could range from a minor abrasion requiring only a plaster to a fatal injury. Depending on the exposure, impact and amount of energy transferred, the area of body involved, and the duration of the energy exchanged, Luck Factor 3 determines how badly the person or persons are about to be injured (Figure 14.1).

DEFINITION

The *severity* of the injury could be defined as: "the degree of injury or illness suffered and the extent of the disability of the person that was injured."

The Auditor's Guide, NOSA (1995) defines an *injury* as: "any damage to people usually caused by a single short exposure of an individual to the high energy exchange between two or more sources of energy" (p. 20).

NOSA further defines a *disease* as: "any damage to people most usually caused by the frequent, continuous, extensive or long exposure of an individual to one or more sources of energy." Its definition of *disability* is "the inability or limited ability to perform a pre-injury activity in the same manner or within the same range as was possible prior to the damage" (p. 20).

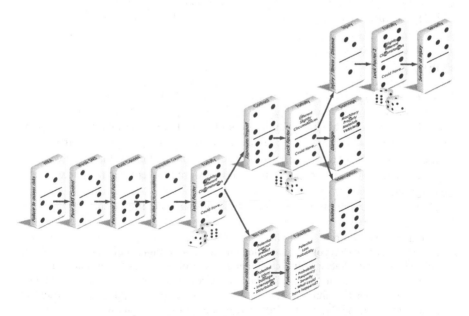

FIGURE 14.1 The severity of an injury is depicted by the 14th domino.

DOI: 10.1201/9781003385943-16

EXTENT OF INJURY

The injury, and more specifically the extent of injury, is what is traditionally linked closest to the terms "safety" and "accident." Most people focus on injuries and the focus is intensified by the severity of the injury. A fatal injury is perhaps the most dreaded experience that management could have, yet the energy exchange that killed the person could have, under slightly different circumstances, inflicted only minor injury or injury less severe than a fatal injury.

SAFETY PERFORMANCE MEASUREMENT

The number of serious injuries is used to gauge safety management performance and safety success. Because so much focus and emphasis are put on the severity of injury, safety practitioners throughout the world have often been inclined to focus their attention on the injured person in an effort to ensure that the injury is not regarded as "measurable" or "reportable."

All Injuries Lose Time

Certain injuries must be reported by law and these injuries are also selected by their severity. The most popular significant injury is the "lost-time injury," or the more modern version "the disabling-injury." All injuries are in fact "lost-time injuries" but what the term intends is that the worker loses a complete work shift, other than the shift on which he or she was injured, as a result of the severity of the injury. The *disabling-injury* definition is similar to the *lost-time* definition except it is more explicit and states that the injured employee is *temporarily partially or totally disabled for a shift other than the shift on which he or she was injured.*

SURVEY

Dave Johnson reported that in the 15th Annual White Paper survey conducted by the *Industrial Safety and Hygiene News* (December 1998b), management was asked how safety and health specialists could contribute to their companies. Ninety one percent answered by saying, "reduced injuries and illnesses." A further 76% wanted reduced Worker's Compensation costs. None indicated that they wanted a reduction in the risks but rather indicated they wanted a reduction in the degree of consequence.

Heinrich (1959) emphasized that accidents are what are important:

> Accidents - not injuries - the point of attack. Accident prevention is too frequently based upon an analysis of the causes leading to a major injury. This situation exists, for the most part, because of a misunderstanding of what an accident really is. In reality, where the term "accident" and "injury" are so merged, it is assumed that no accident is of serious importance unless it produces a serious injury. Yet, thousands of accidents having the potentiality of producing serious injuries do not so result.
>
> *(p. 28)*

SAFETY MEASUREMENT

Few organizations measure their "safety" other than by the number of days that they have managed to work without a lost-time injury or by the number of injuries per million workhours worked. These are all measures of consequence. If the degree of consequence is fortuitous, as explained in the previous chapter, what is really being measured is luck and not control.

Some types of measurement, based on injuries are:

1. Days without a fatality
2. Fatality rates
3. Lost-time-injury frequency rates
4. Disabling-injury incidence rates
5. Injury severity rates
6. Total injury rates
7. Reportable-injury rates
8. Injury and severity indexes

All of these measurements are based on *degree* of injury and do not necessarily tell an organization how good its safety controls are.

Dan Petersen (1998) explains:

> In the early days of safety, accident (injury) measures, such as the number of accidents, frequency rates, severity rates and dollar costs, were used to measure progress (of a corporation, location, or unit). Practitioners felt comfortable using those measures even though they did not offer much help – these measures did not reveal whether the safety system was working; diagnose what was/was not working; nor indicate whether the system was in or out of control.
>
> Although it became clear long ago that these measures offered little help, they continue to be used today. Are accident (injury) data useful for anything? It would save much time, effort, and money to simply answer "no," and focus on measures with meaning.
>
> *(p. 37)*

INJURY REPORTING REQUIREMENTS

Health and safety legislation around the world also require injuries, based on the extent of the injury, to be recorded and reported. Different countries have different rules and criteria for reporting and therefore any attempt to compare reportable, recordable, or lost-time injuries among countries is virtually impossible. The only accurate comparison is the fatality rate, which is the only injury rate that cannot be modified, adjusted, or slanted. The rules and extent of a fatality are universal.

Safety Performance

A low injury rate does not necessarily mean good safety performance. An injury rate is not safety performance but rather safety *failure*. Safety audits of the health and safety management system (SMS), which measure and compare control mechanisms introduced to prevent accidental losses, should rather be used as a gauge of safety performance or safety control.

Even today, companies issue awards, plaques, stickers, and prizes for individuals and groups being injury free for certain periods. When asked what these individuals or groups did differently to remain injury free, the answer was that they did nothing. They were more than likely injury free as a result of luck and not control and therefore, these awards were being awarded for their having been lucky.

As Tarrants (1980) said:

> No doubt the most reliable index we have is the number of deaths. One simply cannot hide the body under the rug and forget about it. As the injury decreases in severity, it becomes progressively easier to ignore it or to remove it from the "reportable" or "recordable" category. For example, when pressure is put on the supervisor to cut down on his first aid cases, he may tell his employees not to report their minor injuries to the dispensary, but to see him for some antiseptic and an adhesive bandage.
>
> None of the measures of safety performance commonly used is acceptable as a valid means of identifying our internal accident problem. It is important that one is not lulled into believing that present measures are really descriptive of the actual level of safety effectiveness within an organization.
>
> *(pp. 39–40)*

Simonds and Grimaldi (1963) feel that they are good measures and are useful for comparison:

> *Frequency* and *severity* are accepted standards by which a company can appraise its industrial injury record and set goals for achievement. Very roughly, these terms refer, respectively, to the *relative* frequency of occurrence of major injuries, on the one hand, and the total days lost, plus time charge for deaths and permanent impairments resulting from major injuries, on the other hand.
>
> *(p. 31)*

Annual reports as well as monthly reports of large industrial and mining undertakings now frequently include impressive bar charts and graphs showing the 12-month running disabling-injury incidence rate or lost-time-injury frequency rate. Used as the sole gauge of safety performance, they are unreliable indicators of safety control.

Using the injury-severity rate, which is the number of shifts lost as a result of injuries (which is also dependent on the circumstances of the accident that caused the injury), is not a reliable measure of good safety controls either. Dan Petersen explains (1978):

> Safety professionals for years have been attacking frequency in the belief that severity would be reduced as a by-product. As a result, our frequency rates nationwide have been reduced much more than our severity rates have. One state reported a 33% reduction in all accidents between 1965 and 1975, while during the same period the number of permanent partial disability injuries actually increased.
>
> *(p. 19)*

WHAT IS A LOST-TIME INJURY?

In defining what an organization regards as a lost-time injury, a disabling injury or even a work-related injury or illness leads to increased confusion. This leads to further inaccurate reporting. The definitions differ from organization to organization and from country to country. Comparing injury rates based on different sets of rules gives a totally inaccurate picture.

ANSI

The American National Standards Institute (ANSI) gives the most comprehensive and rigid definitions of injuries in its standard on recording and measuring work injury experience (ANZI Z16.1 and Z16.2). Even with these guidelines, Workers' Compensation rules differ from state to state and country to country, and some organizations accept a compensable injury as a work injury even though it may have occurred off the premises and was not arising out of and during the course of normal employment. These differences once again detract from the accuracy of defining the degree of injury.

GROUP STUDY

Five teams, consisting of five employees each, were asked what a disabling-injury incidence rate of 8 meant. Only one of the teams got the answer close to a plausible explanation of what the magical figure of 8 meant. They answered by saying, "For every 200,000 workhours there are eight disabling injuries." This was the closest answer received from the teams. Past experience has also shown that at least 90% of the workforce do not understand figures quoted depicting disabling-injury incidence rates or lost-time-injury frequency rates. Most of the teams questioned tried to recall the formula and one team actually stated, "The DiiR is measured by incidents or accidents occurring in a specific time frame, total workhours worked divided into the reportable injuries for the period."

PERCENTAGE OF THE WORKFORCE INJURED

The main reason for using the disabling-injury incidence rate instead of the lost-time-injury frequency rate is that the disabling-injury incidence rate is based on 100 workers working for a year. This is equivalent to a percentage of the workforce injured (disabling injuries) per year. This is a measurement with which the workforce can associate. The lost-time-injury frequency rate, on the other hand, is calculated around a million workhours, or 500 people working for a year, which is more difficult to comprehend than percentage of workforce injured.

MAGICAL NUMBER

From discussions, interviews, and tests with employees in numerous industries over the last 35 years, it is clear that not only is the measurement of safety suspect, but it is also *the* magical number. These are displayed on safety publicity boards around the companies and are not understood. This is perhaps a reason that the modern trend is to display the number of days worked without suffering disabling injury. This is a more understandable indicator for the general workforce.

MINOR INJURIES

Since Heinrich's first accident ratio was published, other researchers show that a large number of minor injuries are experienced in comparison with the number of serious or disabling injuries. The first Heinrich ratio estimated that there were 29 injuries for every one serious injury. The Bird and Germain Ratios produced in 1966 predicted ten minor injuries for every serious injury. If an organization uses the injury at the peak of the triangle to measure its safety performance, surely the ten minor injuries should also be taken into account. But for the luck factor, they could also have been disabling injuries. The ratios produced by Heinrich, Bird, Tye, and Pierson all indicate that the disabling injury is preceded by numerous less-severe injuries, which, under slightly different circumstances, could have been more serious, thus heightening the argument that measurements based on severity are really measuring a degree of luck.

SAFETY INCENTIVES

The debate about safety incentives continues. Some management theories advocate the recognition of good performance by giving out incentives. Safety management is no different; except that most incentive and awards are based on a measurement compiled using injuries or injury-free periods. These incentives, awards and competitions are referred to as "traditional safety-incentive programs."

Mark A. Friend (1997) says that sometimes these incentive programs do more harm than good:

> The operative word here is "typical." Typical programs reward employees for hiding facts – by encouraging them to not report injuries. Conversely, sound safety incentive programs reward employees who offer input that helps everyone operate more safely.
>
> *(p. 36)*

Safety incentives do not reduce the number of lost-time injuries or whatever other degree of injury is used to measure so-called safety "performance" or "improvement." In many instances they reduce the reporting of these injuries.

SURVEY

In a survey conducted among 23 employees, 13 (57%) agreed that they would not report an injury if they were paid a monthly bonus of $150 for being "injury free." They unanimously agreed that if the injury was so severe that it could no longer be hidden, or that it obviously affected their ability to work, only then would they report the injury.

Judy Prickett (1998) quotes the following example:

> I recently heard of a company that promised $100 cash to all employees if they exceeded their previous record of days without a lost-time incident (injury). Within

days of attaining the record, one of the employees was injured and reported that injury. He was ostracized by his fellow employees and ended up quitting.

(p. 8)

DRIVING INJURIES UNDERGROUND

So, does a safety-incentive scheme really reduce the number of reportable injuries or does it merely drive those injuries underground? If the degree of injury is the sole measure of safety performance, it can easily be modified to achieve whatever number management deems as being reasonable, or a figure that shows an improvement on previous figures. *Industrial Safety and Hygiene News* (September 1998) asked for feedback from readers as to how much pressure safety departments face to under-report injuries and illnesses. This is to avoid OSHA targeting them now that the agency is using employers lost workday rates to compile lists of inspection targets. An anonymous reply was as follows:

> There are many organizations that will never make safety a priority. They believe they are safe because no one has been killed or seriously injured recently. These folks will probably go to great lengths to fudge the numbers. We will always need a strong OSHA to deal with them!

(p. 21)

Another respondent to the same question in the same article starts his reply with:

> Under reporting would depend on to whom the safety department reported. If they were part of a human resources department, then the pressure would not be as great as it would be if they report to a profit center. There is always pressure to lower an injury rate, but when budgets and bonuses are on the line, it can be intense.

(p. 21)

From the above comments it is apparent that there is tremendous pressure to lower injury rates by any means other than to institute proper controls based on risks identified by the risk assessment process. Dr Mike Manning (1998) cautions the safety practitioner as follows:

> I have been tempted to make exceptions to the rule for the incident rate, but I never did. I always required that the injury be recorded by the department that it occurred in. If you start making exceptions for certain cases, the whole intent of the incident rate, as a measuring tool will be lost. Stand firm on this. Some supervisors will attempt to manipulate, cajole, threaten, and whine their way out of accepting their responsibility to their employees, but don't let them.

(p. 36)

HONESTY

Honesty is therefore required for both the recording and reporting of all injuries and more specifically those that eventually meet the criteria as the *attention getter* or measuring injury such as the disabling or lost-time variety. It should be mentioned that all undesired events "lose time" and therefore the use of the expression "lost-time injury" is outdated as all injuries *also* lose time. Industries, management, and safety professionals should start using correct terminology and revert to using the term "disabling injuries" as defined by the ANSI Z16 standard.

If one therefore summarizes the conditions needed to have meaningful injury statistics, they should be honest and unbiased. Even though they are unbiased, are all injuries being reported? If the rules for determining disabling injuries are being followed, whose rules are these and how many exceptions have been made to them?

Disabling-Injury Intent

The intent of the disabling-injury definition was to classify injuries that actually hampered the business process, where the work in progress at the time of the undesired event could not be continued exactly as before, due to the undesired event (accident). The only acceptable definition of a disabling injury is to ask, "What was the person doing at the time of the accident?" And "Can he or she return to continue the same task that they were doing after the accident without losing a complete shift?" If one were to use the above criteria to determine whether an injury is disabling, the disabling-injury incidence rate, I believe, would double.

Dan Petersen (1996) says:

> In the early days of safety, it was simple – we just looked at the statistics; a 0.5 accident (injury) frequency rate meant we were good and that our safety program was effective, whereas a 15.0 frequency rate meant that our program was not effective. It was simple; the frequency rate (of recordables or lost times) told us whether or not everything was working. And this was true until we started looking at the real truth – at the meaninglessness of the number we are using.
>
> *(p. 15)*

The Health and Safety Executive (HSE) of the United Kingdom carried out an extensive safety study via its Accident Prevention Advisory Unit and summarized this as follows:

> Any simple measurement of performance in terms of accident (injury) frequency rates or accident/incident rate is not seen as a reliable guide to the safety performance of an undertaking. The report finds there is no clear correlation between such measurements and the work conditions, in injury potential, or the severity of injuries that have occurred. A need exists for more accurate measurements so that a better assessment can be made of efforts to control foreseeable risks.
>
> *HSE Website (2022)*

CAN HEALTH AND SAFETY MANAGEMENT SYSTEMS HELP?

It has always been believed, and many safety practitioners still believe, that an intensive, management-driven, SMS will reduce the number of injuries. They believe this even though studies such as those done by the *Health and Safety Executive* found that often there is no correlation between the degree of control being exercised by an organization and the number of disabling injuries being experienced. Even with SMSs being regularly audited and acclaimed by external auditors, disasters can occur.

In 1988, the Middelbult Colliery had maintained a five-Star National Occupational Safety Association (NOSA), 5-Star Safety Grading for three years. This entailed maintaining 95% control of 73 crucial health and safety elements and a 12-monthly progressive disabling-injury incidence rate of <1% per annum. Even with all this effort, an underground methane explosion occurred, and 53 miners died instantly. Was this perhaps as a result of the three luck factors that have been discussed? Health and safety systems and control systems can only drive down the *probability of* undesired events occurring, but the actual end result is attributable to the three luck factors. The degree of an injury, as well as the severity of an injury, is fortuitous and therefore using them as sole measures of safety efforts is fruitless.

CASE STUDY

To further emphasize the futility of using the degree of injury as a measure of safety or non-safety, the following case study is given.

An underground miner was working on the 65-ft level and entered across Section 16. As he reached up to hook his fall-protection line to the static line, he slipped and fell through the rails. He fell 120 ft (36 m) to the ground level, fracturing his left ankle. Further injury was averted by the build-up of material on the side of the uprights that slowed his descent. This was not recorded as a lost-time or disabling injury as the employee was back at work the following morning even though he was walking with the aid of crutches and his ankle had been tightly bandaged. It was recorded as a minor injury.

IDENTICAL ACCIDENT

The irony is that an identical accident occurred some two years previously, where an employee fell the same distance and was killed. That accident was reported as a fatal accident, and besides shocking the workforce and the local community, resulted in an investigation by the local sheriff, the state safety organizations as well as federal organizations. The only difference in the two scenarios is Luck Factor 3, which, simply said, means that *luck* determines the severity of the injury.

CONCLUSION

The degree of injury is as a result of certain factors referred to in this publication as Luck Factors 1, 2, and 3. As a result of three luck factors, the severity of an injury is therefore a poor gauge of either success or failure in safety performance.

Safety incentives over the years have taught employees, supervision management, and indeed, the safety profession, how to manipulate the figures, bend the rules, and produce low injury rates to justify their existence or their supervisory skills. Regulatory authorities have added to this dilemma by basing harsh inspections on companies with injury rates higher than the national injury rates for that type of industry. In other words, the honest organization that methodically reports and records the number and severity of its injuries is likely to be punished by being subject to an inspection and subsequent citations. Employees themselves are encouraged not to report injuries. If no injuries of a serious nature occur, management turns a blind eye and thinks that all is well and that it has control. Many a management team has been absolutely staggered by a fatality. In many instances, the potential for a fatality had been present all along. The same event probably occurred without severe result many times in the past.

Dave Johnson (1998a), editor of *Industrial Safety and Hygiene,* quotes Henry Lick, who is a committee member and manager of industrial hygiene for the Ford Motor Company:

> We don't get progress until we kill people. It is too bad progress has to be measured in blood. Whole generations go by without problems being solved.
>
> *(p. 21)*

15 Costs of Accidental Loss

Although healthy and safe workplace conditions can be justified on a financial basis, many employers prefer to justify them on the basic principle that it is the right thing to do. In discussing safety in industrial and mining operations, it has often been stated that the cost of adequate health and safety measures would be prohibitive and that organizations can't afford it. The answer to that is quite simple and direct and is: "If you can't afford safety, you can't afford to be in business."

ACCIDENT COSTS

The total economic cost of fatal and nonfatal preventable injury-related incidents in the US during 2020 was $1,158.4 billion. This includes employers' uninsured costs, vehicle damage, fire costs, wage and productivity loss, and medical and administrative expenses. Motor vehicle-related injuries cost $473 billion, work-related injuries, $163 billion, and public injuries, $166 billion (Figure 15.1).

INCREASED PREMIUMS

The end result of an accident can always be translated into costs. Whether the event results in injury, disease, and damage to machinery, property or materials or business

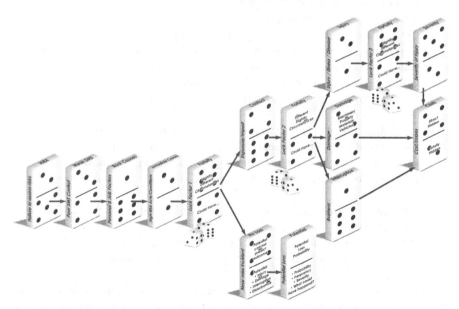

FIGURE 15.1 The Cause, Effect and Control of Accidental Loss (CECAL) accident domino sequence showing the resultant cost of accidents.

DOI: 10.1201/9781003385943-17

interruption, they all cost the organization money. Traditionally, these costs have been tolerated as the cost of doing business and have not received management's full attention.

Workers' Compensation normally covers the direct costs of the worker injured in an accident. These premiums can be increased as a result of injury costs and these increases could remain for three-year cycles or periods. Most of the accidental costs are hidden costs or indirect costs. These are difficult to calculate and are often ignored because they do not create an immediate financial drain on the organization.

INCIDENTAL COSTS

In addition to the obvious costs, there are also the so-called incidental costs of accidents. These incidental costs of accidents have been estimated to be up to five times as great as the actual costs. There have been innumerable studies made, discussions held, articles written, and arguments presented about that figure. No one has disputed the concept that there are indirect, incidental, or hidden costs surrounding accidents. Most tend to agree with that. There is tremendous disagreement, however, about how much those costs might amount to.

The following is a list of 11 types of hidden accident costs, which exclude compensation and liability claims, medical and hospital costs, insurance premiums, and costs of lost time, except when actually paid by the employer without reimbursement. These 11 main hidden costs are:

1. Cost of lost time of injured employee
2. Cost of time lost by foremen, supervisors, or other executives
3. Cost of time lost by other employees who stopped work
4. Cost of time spent on the case by first aid attendant and hospital department staff who were not paid for by the insurance carrier
5. Cost to employer under employee welfare and benefit systems
6. Cost due to damage to the machine, tools, or other property, or to the spoilage of material
7. Incidental costs due to interference with production, failure to fill orders on time, loss of bonuses, payment of forfeits, and other similar causes
8. Cost to employer in continuing the wages of the injured employee in full
9. Costs due to the loss of profit on the injured employee's productivity and on idle machines
10. Costs that occur in consequence of the excitement and worker morale due to the accident
11. Overhead costs per injured employee (light, heat, rent, and other such items, which continue while the injured employee is not productive)

The list of costs as a result of an accident can be a lot longer depending on the circumstances. Safety pioneers established that there is a definite ratio between the insured (direct) and uninsured (indirect) costs of accidents and *conservatively* put the ratio at 1:5.

TOTALLY HIDDEN COSTS

A further category can be added to the costs of an accident. Those costs are termed the totally hidden costs. These costs are effects that cannot be costed out accurately and include such intangible costs as the cost of pain and suffering of the injured employee, the disruption of the victim's family life, the lowered standard of living due to his or her pay packet being reduced substantially, as well as the accompanying stress and anxiety caused by the accident. The loss to the community and the organization is also difficult to compute.

ICEBERG EFFECT

McKinnon (1995) refers to the iceberg effect:

> The iceberg effect is where the majority of accident costs are hidden below the waterline. The hidden costs or indirect costs of accidents include items such as damage, loss of time and production and business interruptions. Experts have put these costs at anywhere from 5 to 50 times more than the direct costs. Irrespective of what figures are placed there is definitely a higher cost of hidden costs than direct costs.
>
> All costs are recuperated from the profits of a company. Peter Drucker, the well-known management consultant and author, states that it would be better to *minimize* losses than to constantly endeavor to *maximize* profits. Management is often aware of the efforts to maximize profits by improving quality and productivity but is not always aware of losses occurring as a result of accidents. The prevention of all downgrading accidents is a good investment and can improve the company's bottom line substantially.
>
> *(p. 31)*

In summarizing the advantage of safety control, experts agree that if you save a dollar in accident costs you add a dollar to profit. All agree that if you add this to the overriding human element you have the best of both worlds: protection of profits, process, property, and people. That is why it is so essential to understand and use the accident cause and effect sequence.

The ratio between the injury and illness costs, and the ledger costs of property damage could be anywhere between 5 to 1 and 50 to 1. The uninsured, miscellaneous costs could be as high as 1–3 dollars per every dollar direct cost (Figure 15.2).

EXAMPLE

In motivating the amount of profit consumed by accident costs, an excellent example is, if an organization's profit margin is 5%, it would have to make sales of $500,000 to pay for $25,000 worth of losses. With a 1% margin, $10,000,000 of sales will be necessary to pay for $100,000 of the costs involved with accidents.

MAIN MOTIVATION

For years, safety practitioners and safety organizations have promoted the humanitarian aspect of safety and have endeavored to reduce injury-producing accidents. The humanitarian approach is all very well, but this is not what motivates management.

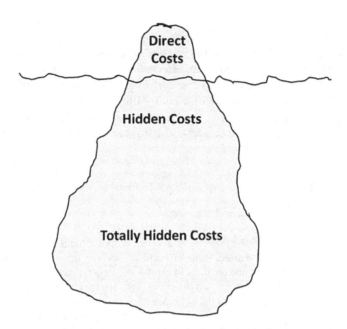

FIGURE 15.2 The Iceberg effect which shows more hidden cost below the waterline.

PROFIT DRIVEN

Management is in business to make a profit and give shareholders a return on their investment. Profits and the bottom line are what an organization is all about, and what management at all levels is ultimately held accountable for. If the costs of accidents due to poor control can be brought to management's attention, the necessary actions and support will be forthcoming. Yet so many organizations concentrate on the numbers of injuries, and endeavor by all means to reduce these numbers, that they lose sight of their greatest management attention getter, *the total cost of accidents.*

One of the major objectives of accident prevention work is to reduce production or operating costs for the sake of profit. While second to the prevention of human injury, since that is even more directly important as a human value, cost reduction broadens the basis for safety work. Cost reduction provides a direct purpose for preventing all kinds of accidents, accidents that cause injury as well as other undesired events. Cost reduction brings into focus the losses from property damage and interference with production as well as those from injury accidents.

Individuals, promoters, and organizations geared to prevent accidental loss will continue to be unsuccessful unless they start to justify their recommendations on sound financial basis.

Compliance Magazine (CM) 1998, in an interview with Jerry Scannell, president, National Safety Council (October 1998), he was asked the question, "How can a company achieve excellence in safety?" Scannell replied, "Success will not happen overnight. A company must hold safety as a core value right up there with making a profit."

In answering the question, "How are you convincing management to promote safety within their companies," Scannell answered, "If you are trying to convince your company's chief executive officer that safety is important, you must show that safety offers financial rewards. 'Cut costs' are two words that corporate America is using today."

In emphasizing the importance of hidden costs, Dan Petersen (1978) says:

> Hidden costs are real. Many people today believe that the dollar is a far better measuring stick than any other in safety, and many companies are beginning to utilize it effectively.
>
> *(p. 50)*

It should be remembered that if the severity of injury is as a result of Luck Factor 3, then the subsequent direct costs of the injury or disease are also largely fortuitous and therefore although the dollar cost is a useful measurement, it, like injury rates, is highly subjective.

FINES

Most organizations are required by legislation to maintain a health and safety management system (SMS), which gives certain controls of risks that have been identified. Although these controls, checks, and balances do cost money, they should be viewed as an investment, as accidental losses could be far more expensive. Non-compliance to health and safety legislation could also cost a company a staggering amount of money.

COST OF NON-COMPLIANCE

The cost of non-compliance to health and safety laws and regulations can cost an organization a great deal, even before the occurrence of an accident. OSHA lists the top ten violations of 2020 and the fines for non-compliance.

1. Fall Protection Violations – penalties: $22,767,761.
2. Hazard Communication – penalties: $2,719,806.
3. Respiratory Protection – penalties: $3,172,334.
4. Scaffolding in Construction – penalties: $5,062,635.
5. Ladders in Construction – penalties: $4,785,679.
6. Hazardous Energy (Lockout/Tagout) – penalties: $9,633,595.
7. Powered Industrial Trucks – penalties: $3,765,006.
8. Fall Protection—Training – penalties: $2,051,699.
9. Eye and Face Protection – penalties: $3,521,988.
10. Machinery and Machine Guarding – penalties: $6,932,297 (OSHA Website, 2022)

HIGHEST EVER

MSHA issued the largest fine in its history, $10.8 billion against the former Massey Energy Co. in connection to the 2010 Upper Big Branch Mine explosion. In its

December 6 fatal accident investigation report, MSHA attributed the root cause of the disaster to a corporate culture that valued production over safety.

On April 5, 2010, a massive explosion in the Upper Big Branch Mine, which was operated by Performance Coal Co., a subsidiary of Massey Energy Co., killed 29 miners and injured two others. MSHA has now issued Massey and PCC 369 citations and orders, including an unprecedented 21 flagrant violations, which carry the most serious civil penalties available under the law.

"The results of the investigation led to the conclusion that PCC/Massey promoted and enforced a workplace culture that valued production over safety and broke the law as they endangered the lives of their miners," said Secretary of Labor Hilda L. Solis. "By issuing the largest fine in MSHA's history, I hope to send a strong message that the safety of miners must come first."

MSHA announced its report findings and fines following the $209 million settlement and non-prosecution agreement reached December 6 among the U.S. Attorney's Office for the Southern District of West Virginia, the U.S. Department of Justice, Alpha Natural Resources Inc., and Alpha Appalachia Holdings Inc., formerly known as Massey Energy Co (MSHA Website, 2022).

MAJOR ACCIDENTS

The Chernobyl nuclear accident cost around US$200 billion. The loss of the space shuttle Columbia cost US$18 million on direct costs of the accident investigation board and $112 million in supporting the board's investigation. The Prestige oil spill cost US$12 billion. The oil spill resulting from the Exxon Valdez disaster cost US$2.5 billion, and that of the Deepwater Horizon, US$61.6 billion. The Bhopal accident cost US$470 million in settlements alone.

COST BENEFIT

An effective approach to the safety problem, which is supported by many other practitioners, is to do a cost benefit analysis during the evaluation phase of a risk assessment. Thorough risk assessment will indicate to management where best to spend its money for the largest return. Traditionally, a number of safety programs have channeled money into activities that have not necessarily tackled the root cause of the accident problems.

When various risk reduction approaches are being considered, one basis for decision is to examine the cost of each, versus the benefit of its application. The alternate measures that produced the greatest benefits for the least cost are normally selected. One problem with this approach is the difficulty in assessing dollar costs of such intangible accident consequences such as pain, suffering, and loss of life.

LIFE'S VALUE

Injury Facts 2020 give the cost of a US work-related fatality at $1,290,000 and a disabling injury between $37,000 and $43,000 depending on employer costs.

One of the largest penalties paid by a South African division of an international organization was when, in 1994, four chemical firms agreed to pay 20 South African workers ZAR9,4 million (US$1.5 million) in damages and costs.

REPUTATION

One of the hidden costs of an accident is the cost a company suffers when its reputation is tarnished as a result of a serious accident or series of accidents. On the day following the Vaal Reefs disaster in South Africa on 17 May 1995, the shares of Anglo-American Gold Division, the owners of Vaal Reefs, fell ZAR16 (US$3) per share in one day.

SEVERE REPERCUSSIONS

Even ignoring the intangible costs, an accident that results in a fatality normally has severe repercussions on the organization. In one particular fatal investigation, the costs were broken up into injury, property damage, business interruption, and total hidden costs. The summary of this accident report was "the losses of the accident are listed under the headings of injury, property damage, interruptions, and total hidden costs. Total cost is estimated to be far in excess of US$1,000,000."

Even property-damage accidents can hamper the production of an organization and one particular property-damage accident, which resulted in a seven-day period to repair the damage, resulted in a loss of 5,000 units being produced.

WHITE PAPER

A survey was conducted which asked management to list the most important contribution health and safety personnel can make to their companies. Only 37% of the managers listed, "Document the financial impacts of safety activity." When asked what skills and knowledge are important for a safety professional, 17% responded, "The ability to document dollar savings of safety activity." It was also interesting to note from the survey that almost half (44%) of the managers surveyed believed accidents will happen, regardless of safety efforts.

TOTAL COST OF RISK

There are three main areas where costs play a role in safety. The first is the cost of the end result of undesired events such as injury and property-damage accidents. As discussed, these costs could have as many as three different tiers and can range from tangible to intangible and from direct to indirect costs. They are losses to the organization, nevertheless. The second cost of risks is the cost to insure equipment, plant, product, and personnel. These are the Workers' Compensation costs, insurance costs, and such like. The third cost is that involved in reducing, containing, and minimizing the risks, which manifests in loss and cost-producing events. This is the cost of the SMS and personnel.

SUMMARY

The CECAL sequence always ends up with costs as the last effect. Once the event results in an exposure, impact, or contact with a source of energy, the losses could be due to injury, property damage, business interruption, or a combination of all three as already discussed. The costs of assessing and controlling the risk have proved to be less expensive than the cost of the consequence of the event. The benefit of reducing and controlling the risk is also the avoidance of heavy legal fines for non-compliance to legal safety standards. The costs of safety controls are therefore a good investment and as one safety professional put it, "If you think safety is expensive, try an accident!"

Part 3

Accidental Loss – The Control

16 Safety Management Functions

INTRODUCTION

If the hazards and consequent risks arising out of a business have not been identified and assessed, they cannot be managed or controlled. This creates lack of, or poor management control, which is an indication of; a non-existent structured health and safety management system (SMS); a weak system; or non-compliance with the requirements of the SMS. It is this failure that triggers off the sequence of events that culminate in accidental loss.

As Lester A. Hudson said:

> There is a great tendency – human tendency – for management to rationalize after experiencing a human tragedy. It is always so much easier to find the "careless acts" on the part of an injured employee which precipitated the accident, but an enlightened management will not hesitate to look beyond the "unsafe act" on the part of an employee and to consider it as a symptom of lack of management control (p. 2).

MANAGEMENT LEADERSHIP

The health and safety of employees at a workplace is the ultimate responsibility of the management of the organization. Even though it is generally accepted that all share a role in safety, the ultimate accountability lies heavier with all levels of the leadership. With this in mind a SMS can only be successful if initiated, led, and supported by line management.

The four key elements for safety performance are:

1. Top management commitment
2. A humanistic approach towards workers
3. One-on-one contact
4. Use of positive reinforcement

The following five SMS elements are vital components for an effective SMS:

1. Management commitment
2. Employee involvement
3. Worksite analysis
4. Hazard identification (HIRA) prevention and control
5. Health and safety training

DOI: 10.1201/9781003385943-19

WHAT IS A MANAGER?

A manager is anyone who uses management skills or holds the organizational title of "manager," and who gets things done through other people. Management can also refer to the person or people who perform the act(s) of management.

BASIC MANAGEMENT FUNCTIONS

Over the years it has generally been accepted that a manager's main functions are:

- Planning
- Organizing
- Leading or Directing
- Controlling

All these functions entail the management of employees, materials, machinery, and processes, and they also apply to health and safety management.

These management functions also refer to the activities of a manager appertaining to the health and safety of his or her resources, namely, people, material, equipment, and the work environment.

If the four basic functions of management are integrated into a manager's normal functions, it could provide for better management, leadership, and involvement in the SMS and its elements (Figure 16.1).

Safety Planning	Safety Organizing	Safety Leading	Safety Controlling
Safety forecasting	Safety and health integration	Making safety decisions	Identifying hazards
Setting objectives	Safety delegation	Communicating safety	Identifying the work to be done
Setting policies	Creating safety relationships	Motivating for safety	Set standards of measurement
Safety Programing	Safety authority	Appointing employees	Set standards of accountability
Safety Scheduling	Safety responsibility	Developing employees	Measure against the standard
Safety Budgeting	Safety accountability		Evaluate conformance
Establishing safety procedures			Corrective action
			Commend compliance

FIGURE 16.1   The four basic safety functions of management.

SAFETY PLANNING

Safety management planning is the process of developing an organization's health and safety goals, and creating a realistic, detailed plan of action to develop and maintain a SMS to meet those goals. Safety management planning takes into consideration short-and long-term organizational strategies.

Safety planning is what a manager does to predetermine the occurrence and consequences of accidents, and to determine action to be taken to prevent these downgrading events occurring.

THE FUNCTIONS OF SAFETY PLANNING

Safety Forecasting

Safety forecasting is the activity a manager carries out to estimate the probability, frequency, and severity of accidents and near-miss incidents that may occur in a future time span. This is usually done by means of hazard identification, risk assessment, critical task identification, and task risk assessment. Accident and near-miss incident analysis can predict what losses could have occurred if the event had not been identified and the root causes eliminated.

Setting Safety Objectives

Setting safety objectives is when a manager determines what safety results they desire. This would include involve setting key safety performance indicators. These should be leading, manageable, achievable, measurable, actions and activities. These indicators should not be based on injury rates. Objectives based on injuries should be avoided as the degree of injury is often fortuitous.

An example of a leading indicator could be the number of safety inspections to be carried out each month, or the number of critical task procedures that need to be reviewed annually.

Another objective for the organization could be the number of employees trained in health and safety each month, or the number of hazards that have been identified, reported, and rectified.

Setting Safety Policies

Setting safety policies is when a manager develops standing safety decisions applicable to repetitive problems which may affect the health and safety of the organization. The health and safety policy statement is the leading document in the safety drive and should be established by senior management. This function would also entail having standards (policies) written for each element of the SMS.

Safety Programming

Safety programming is establishing the priority and following order of the safety action steps that must be taken to reach the safety objective. An example would be determining what percentage of reported high-risk behavior and conditions and other hazards must be rectified within certain time limits. Another important activity is setting standards for communicating this information down to the workforce. Actions should also be planned in relation to the nature of the hazards to be eliminated.

Safety Scheduling

Safety scheduling is when a manager establishes time frames for the implementation of the elements, processes, and programs within the SMS. Schedules need to be set for risk assessments to be done, safety inspections to be carried out, and health and safety committee meetings to be held, among other activities. In introducing and maintaining a SMS, a schedule would be determined for the introduction phase of the system, the training phase, as well as the follow-up and review of the results, successes, and failures of the system.

Safety Budgeting

Safety budgeting is allocating financial and other resources necessary to achieve the health and safety objectives of the organization. A budget allocation may be required for the health and safety department. Funds should be allocated for training. Mechanical or structural repairs or modifications may be needed to eliminate hazards reported through the hazard reporting system, and these expenses must also be budgeted for. Safety equipment and annual training also require budgets.

Establishing Safety Procedures

Establishing correct safety procedures is when a manager analyzes certain tasks and ensures that safe work procedures for performing hazardous work safely are written. Using a task risk assessment process, the tasks can be risk-ranked, and the critical tasks identified. This will help prioritize the writing of procedures.

Numerous procedures and processes within the SMS need to be developed and established along with the appropriate training. These could include accident investigation procedures, emergency evacuation procedures as well as hazard and risk assessment procedures.

Specific health and safety system processes, procedures, and methods must be developed, and designated employees trained and appointed to coordinate these activities.

SAFETY ORGANIZING

Safety organizing is the function a manager carries out to arrange safety work to be done most effectively by the right people. This would involve allocating persons to coordinate the SMS, to conduct inspections, do risk assessments, develop training, etc.

THE FUNCTIONS OF SAFETY ORGANIZING

The following are the basic functions of safety organizing.

Integrating Safety into the Organization

Integrating safety into the organization is the work a manager carries out to allocate safety work to be performed by the various levels within the organizational structure. This would include the responsibility to manage and coordinate activities within the SMS. The SMS processes, procedures, and actions should be integrated into the day-to-day business of the organization and not be regarded as stand-alone activities divorced from normal operations. The more the safety management system (SMS) is integrated into the organization, the better the safety culture becomes.

Safety Delegation

Safety delegation is what a manager does to give safety authority to his subordinates and entrust safety responsibility, while at the same time creating accountability for safety achievements. Managers should be given safety authority, safety responsibility and be held accountable for safety actions. All employees should be given the authority to report hazards, accidents, and near-miss events and to participate in the SMS activities. Once problem areas have been identified, managers are then held accountable to rectify the deviations depending on their area of control and level of authority and responsibility.

The level of accountability should be the same as the level of authority that an employee has. Making employees "responsible for safety" is too broad a statement and makes employees accountable for things over which they have no authority. Employees should only be held accountable for their own safety and things over which they have authority.

Creating Safety Relationships

Creating safety relationships is done by a manager to ensure that safety work is carried out by the team with utmost co-operation and interaction amongst team members. All levels within the organization should have some responsibility with regard to health and safety. The SMS must be owned by all levels and should not be seen as a bargaining tool or a system to gain personal benefits or demands. The SMS requires participation, support, and action from all levels within the organization, and cannot be left as one person's or one department's or one manager's, responsibility.

Allocating Safety Authority

Safety authority is the total influence, rights, and ability of the position, to command and demand safety. It is the right or power assigned to a manager to achieve organizational objectives. Management has ultimate safety authority therefore is the only echelon that can effectively implement and maintain an effective SMS. Leadership has the authority to demand compliance to the standards within the SMS. They also have the authority to take necessary remedial actions should deviations from standards occur.

Assigning Safety Responsibility

Safety responsibility is the obligation to carry forward an assigned task to a successful conclusion without accidental loss occurring. It is also the obligation entrusted to the possessor, for the proper custody, care, and safekeeping of property or supervision of individuals. It is the safety function allocated to a position. It is the duty and function demanded by the position within the organization.

Responsibility lies with all levels of management as well as with employees. The higher the management position, the higher the degree of safety responsibility. One cannot be held accountable for something over which one has no authority or responsibility. The level of safety accountability is also apportioned to the level of safety authority. Job descriptions are vital management tools and should clearly define the safety authority, responsibility, and accountability for all jobs and all levels within the organization. The organization's SMS health and safety standards must clearly define these relationships for the system to be a success.

Creating Safety Accountability

Creating safety accountability is when a manager is under obligation to ensure that safety responsibility and authority is used to achieve both safety and legal health and safety standards. Accountability is to be answerable for actions and deeds performed under safety responsibility. Employees too have safety accountabilities, but in proportion to their safety authority.

Leadership is accountable to implement and manage the SMS, and to provide the necessary infrastructure and training to enable the system to work. Employees should be held accountable for participating in the SMS as required.

Management at all levels is then held accountable to rectify the problems identified by processes within the SMS and to ensure that the high-risk behaviors or conditions, and other deviations, highlighted by the system are rectified and do not recur.

SAFETY LEADING

Safety leading is what a manager does to ensure that employees act and work in a safe manner. It entails taking the lead in safety matters, making safety decisions, and always setting the safety example. Visible felt leadership is an important aspect of leadership. This is one of the most important management functions in implementing and maintaining a SMS.

THE FUNCTIONS OF SAFETY LEADING (DIRECTING)

Making Safety Decisions

Making safety decisions is when a manager makes a decision based on safety facts presented to him or her. Based on the philosophy of accident prevention, managers should take the decisions necessary to implement and support the functions and activities within the SMS.

Safety Communicating

Safety communicating is what a manager does to give and get understanding on health and safety matters. Management and employees' expectations concerning participation in the SMS must be clearly communicated. Standards must be set and communicated to all concerning the requirements of their role in the ongoing SMS activities. Ongoing safety committee meetings and toolbox talks help facilitate health and safety communication. Safety newsletters and notices posted on the safety notice boards help facilitate communication concerning health and safety matters.

Motivating for Safety

Motivating for safety is the function a manager performs to lead, encourage, and enthuse employees to take action for safety. Acknowledging the reporting of hazards by employees is vital to the success of the SMS, and this feedback acts as a reward for such reporting. The power of this small gesture of thanks, acknowledgment, and recognition cannot be overstated.

Positive behavior reinforcement is the key to the success of any SMS. Of all the functions carried out by leadership, it is likely to have the most effect on the success of the system. It demands playing the ball and not the man. It requires managers to fix the problem and not the person. It forces leaders to deviate from traditional management styles when dealing with issues normally calling for disciplinary measures. It will challenge leadership at all levels but will help create more positive leadership across the organization. A good manager catches employees doing things right. Rewarding positive safety actions will have more effect than punishing unsafe behavior.

Appointing Employees

Appointing employees is a management function where management ensures that the employee is both mentally and physically capable of safely carrying out the work for that position. The correct person in the correct work position will eliminate the risk of errors and lapses occurring which could lead to accidents.

Developing Employees

A manager develops employees by helping them improve their safety knowledge, skills, and attitudes. For example, employees should be selected and trained as health and safety representatives. These representatives can then assist management by conducting regular inspections and reporting hazards to them. Management has to ensure the health and safety staff are up to date with the latest trends in safety and risk management and that there is an ongoing self-development program in place for them. Further studies, as well as membership in local, regional, and national safety and health associations should also be encouraged.

SAFETY CONTROLLING

Safety controlling is the management function of identifying what must be done for health and safety, inspecting to verify completion of work, evaluating, and following-up with safety action. This is the most important safety management function and is an ideal process for developing, implementing, and maintaining a structured SMS. The controlling function has eight steps:

I – Identify hazards and assess the risk.
I – Identify the work to be done to mitigate and control the risks.
S – Set standards of measurement.
S – Set standards of accountability.
M – Measure conformance to standards by inspection.

E – Evaluate conformances and achievements.
C – Correct deviations from standards.
C – Commend compliance.

The management function of *control* will be discussed in depth in the next chapter.

CONCLUSION

The only successful method to combat accidental loss is to implement a risk-based, management-led, and audit-driven SMS which identifies hazards, assesses risk, and introduces risk controls on an ongoing basis. This combination of standards, procedures, processes, and ongoing activities will reduce the probability of an accidental loss occurring. This is managements' prime responsibility and is a management function.

17 Safety Management Control

SAFETY CONTROLLING

Safety controlling is the management function of identifying what must be done for health and safety, setting health and safety standards, inspecting to verify completion of work, evaluating, and following up with safety action. This is the most important safety management function and is vital to the design, implementation and maintenance of a successful health and safety management system (SMS).

RISK-BASED, MANAGEMENT-LED, AUDIT-DRIVEN SMS

Based on risk assessments, a manager lists and schedules the work needed to be done to create a healthy and safe work environment and to eliminate high-risk behavior of employees and high-risk workplace conditions. This would mean the introduction of a suitable risk-based, management-led, and audit-driven, structured SMS based on world's best practice and aligned to the risks of the organization. All SMSs should be based on the nature of the business and be risk-based, management-led, and audit-driven. The management control function has eight steps:

I – Identify hazards and assess the risk.
I – Identify the work to be done to mitigate and control the risks.
S – Set standards of measurement.
S – Set standards of accountability.
M – Measure conformance to standards by inspection.
E – Evaluate conformances and achievements.
C – Correct deviations from standards.
C – Commend compliance.

STEP 1 – IDENTIFY THE HAZARDS AND ASSESS THE RISK

The hazard identification and risk assessment (HIRA) process discussed in Chapter 1 will ensure that an undertaking has identified all the hazards, analyzed the risks, evaluated them, and ascertained which risk control methods to apply. These controls would form the basis of the SMS.

STEP 2 – IDENTIFY THE WORK TO BE DONE TO CONTROL AND MITIGATE RISKS

Once the risks have been assessed, evaluated, and prioritized, it is now management's function to identify what work must be done to ensure the treatment of the

DOI: 10.1201/9781003385943-20

risks. The risk assessment would have identified both physical and behavioral risks. Management can now implement certain control elements under the umbrella of the SMS to reduce the risk as low as is reasonably practical.

The work to be done to reduce risks, in many instances, is very similar to the basic health and safety control activities required by industrial and mining legislation. Examples of the work that may need to be done are the following:

- guarding of all machinery and pinch points
- regular inspections of lifting gear
- hazard identification and risk assessments
- safety induction training for new employees
- providing and maintaining of personal protective equipment (PPE)
- hazardous substance control
- controlling permit required work
- formal accident investigation procedures
- legal injury and disease reporting
- hazardous work procedures and controls
- establishing policies and standards, etc.

STEP 3 – SET STANDARDS OF MEASUREMENT

It has been said that managers get what they want, and should management set health and safety standards, these standards are usually achieved by the organization. Setting standards of measurement clearly indicates how things must be in the work environment. Health and safety standards are measurable management performances. Standards must be in writing and should contain the following headings:

- purpose
- resources
- responsibility
- legal
- general
- monitoring

By setting standards of measurement, management defines the direction in which the organization moves. Should management set a standard for good housekeeping practices, this standard, which is a measurable management performance, is then the way business is done in the future. What gets measured gets done. Standards give the company goals and directions and a definite focus as to the end result. These standards can be of measurement and of performance and ask the questions, "What must the end result be?" and "What must be done by whom and by when?" (Figure 17.1).

Example of a Standard for Fiber Lifting Slings

An extract from a standard for fiber lifting slings reads as follows:

- Natural and synthetic fiber rope slings shall be immediately removed from service if any of the following conditions are present:

FIGURE 17.1 Standards of measurement should be established for all elements of the SMS including fiber slings.

- Abnormal wear.
- Broken or cut fibers.
- Variations in the size or roundness of strands.
- Discoloration or rotting.
- Missing load capacity tag.
- Visible "red" warning strand.
- No welding, grinding, heating or repairs may be done to lifting hooks.
- Multiple leg slings shall have each sling and hook in good working order.
- Slings must be stored on suitable racks, shelves, or hooks, off the ground and in clean, dry areas.
- A load chart detailing the ratings of slings at various angles must be posted where lifting slings are stored.
- In certain applications, lifting may only be done under the supervision and control of a qualified person.
- Before use, all slings will be checked against the checklist (Figure 17.2).

Standards of measurement should be set for all the elements, processes, and programs within the SMS. This would include:

- housekeeping
- stacking and storage
- hygiene monitoring
- environmental conformance

#	FIBER SLING CHECKLIST	YES	NO
1	Is the sling being used the correct one for the job?		
2	Is there any damage, chaffing or defects?		
3	Is there any damaged or lose stitching?		
4	Any damage due to heat? (surface will look shiny)		
5	Is there any damage due to chemicals?		
6	Is there any sign of solar degradation?		
7	Are there any damaged or deformed end fittings?		
8	Is the SWL marked?		

FIGURE 17.2 An example of a fiber sling checklist which forms part of the SMS standard for slings.

- safety committees
- safe work procedures
- risk assessments
- plant inspections, etc.

STEP 4 – SET STANDARDS OF ACCOUNTABILITY

The next step in the controlling process is the setting of standards of accountability. A standard of accountability indicates *who* will do *what* by *when*. Setting standards of accountability asks, "Who must do it and by when?" An example of setting standards of accountability is the follow up action after an accident investigation. The control steps to prevent a recurrence are what need to be done to prevent a recurrence of the accident. This should be followed by making one person responsible for the action, as well as committing that person to a date for completion (Figure 17.3).

AUTHORITY, RESPONSIBILITY AND ACCOUNTABILITY

There is often confusion about authority, responsibility, and accountability and it is opportune to define these concepts here:

- Safety authority is the total influence, rights, and ability of the post to command and demand safety.
- Safety responsibility is the safety function and duty allocated to a post or position.
- Safety accountability is when a manager is under obligation to ensure that safety responsibility and authority are used to achieve both health and safety, and legal safety standards.

Setting safety standards of accountability involves *who* will do *what* and *when*. An example of a standard of accountability is the role played by health and safety representatives. Health and safety representatives have been given the authority to

WHAT	WHO	BY WHEN	COMPLETED
1. Replace the machine guard on the filler machine	Maintenance Team	June 1	
2. Arrange a housekeeping inspection in the filler machine area	Filler area supervisor	June 4	
3. All hazards to be rectified immediately	Filler area supervisor	June 4	
4. Toolbox talks to be held every morning emphasizing the need to keep machine guards in place and to maintain housekeeping standards.	Filler area supervisor All Team Members	July 6	
5. Health and Safety Representative to do ongoing monthly inspections.	Health and Safety Representative David.	Monthly	

FIGURE 17.3 Action steps to prevent a recurrence: a standard of accountability.

inspect their immediate work area using a safety element checklist as a guideline. This inspection is carried out monthly as prescribed by the standard. The standard of accountability is:

1. Who? – The appointed health and safety representative.
2. Will do what? – Will carry out an inspection of his/her work area.
3. When? – This inspection will be carried out on a monthly basis.

One of the many safety myths is that "everybody" is responsible for safety. In actual fact, individuals can only be *responsible* for items and people over whom they have *authority* and can thus be held *accountable* for only those conditions and people over whom they have authority.

STEP 5 – MEASUREMENT AGAINST THE STANDARD

This control function is when management measures what is actually happening in the workplace against the preset standards. To measure successfully, a walkabout inspection must be carried out and employees doing these inspections should be aware of, and familiar with, the standards. One of the greatest failings in most SMS is insufficient or inadequate inspections.

Systems to enable ongoing measurement against standards are part of a SMS and these could include the monthly inspections of local work areas by appointed health and safety representatives. Audit inspections are ideal measurement tools. Critical task observations also allow opportunity for measurement against standards. The setting of standards and constant measuring against those standards immediately identifies strengths and weaknesses of the SMS. Safety personnel should also conduct formal inspections on a regular basis and compare actual SMS processes and procedures with the standards. A checklist should always be used when doing these inspections as it will serve as a constant reminder of what must be measured.

It should be emphasized here that this form of safety management control and the measurement phase of the control process are not merely measuring and comparing injury statistics with other companies or industries. This is a pure management function of measuring whether the organization is living up to the norms agreed to by management and employees in the form of SMS health and safety standards, and the health and safety policy statement.

CRITICAL TASK OBSERVATION

Another form of measurement against standards is critical task observation. This involves observing an employee carrying out a critical task while following the steps of the written safe work procedure. The written safe work procedure is the end result of critical task identification (task risk assessment) and the critical task analysis process. The observation allows for measurement of workers' performance during the critical task against the prescribed performance dictated by the procedure. The procedure is a standard of performance.

STEP 6 – EVALUATE CONFORMANCES TO SMS STANDARDS AND ACHIEVEMENTS

The evaluation process is the quantification of degree of conformance to the standards set. Evaluation of achievement of standards is normally facilitated through the audit process. A SMS should be driven by audits. These regular audits systematically quantify the degree of compliance to standards. They evaluate the management work being done to combat loss. What gets measured gets done, and consequently the evaluation of compliance to safety standards gives an indication of what is being done and what is not being done. The quantification of safety control actions is far more reliable and significant than the measurement of safety consequences, which are largely fortuitous.

EXTERNAL AUDITS

A number of health, safety, and risk-management organizations provide auditing services for clients. This external audit is of tremendous value to any organization as it is totally impartial and conducted by auditors who are thoroughly familiar with the audit protocol. The audit of an organization's safety system is a structured approach to the quantification of safety compliance and adheres to the following sequence:

1. Pre-audit meeting
2. Audit facilities
3. Audit team
4. Physical inspection
5. Compliance audit

6. Systems audit
7. Documentation review
8. Verification of disabling injury incidence rate
9. Management close-out and audit results

Retrospective

The audit results and percentage achievement are normally based on the SMS achievements during the preceding 12 months. Credit is not given for good intentions but rather for action and activities that have been in operation for at least six months.

INTERNAL AUDIT

Internal audits of the entire SMS to evaluate conformance with standards should be carried out every six months. The internal audit system should follow the same guidelines as an external audit and will culminate in a percentage conformance, as well as a breakdown of conformance against standards for each element.

Figure 17.4 shows the SMS section elements, with the percentage conformance, which was evaluated during the audit. This immediately gives management an indication as to the strengths and weaknesses of the SMS processes. It also indicates where conformance and non-conformances to health and safety standards lie. Each element of the SMS that does not score 100% indicates a weakness in the element, indicating that the standards set for that element have not been fully met. This indicates to management which elements of the SMS require action.

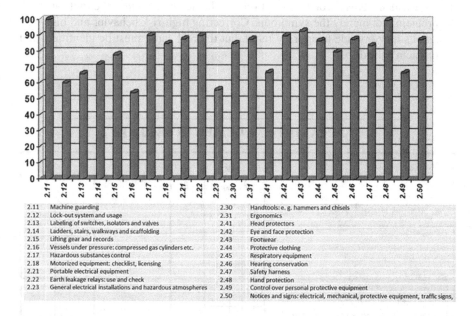

2.11	Machine guarding	2.30	Handtools: e. g. hammers and chisels
2.12	Lock-out system and usage	2.31	Ergonomics
2.13	Labeling of switches, isolators and valves	2.41	Head protectors
2.14	Ladders, stairs, walkways and scaffolding	2.42	Eye and face protection
2.15	Lifting gear and records	2.43	Footwear
2.16	Vessels under pressure: compressed gas cylinders etc.	2.44	Protective clothing
2.17	Hazardous substances control	2.45	Respiratory equipment
2.18	Motorized equipment: checklist, licensing	2.46	Hearing conservation
2.21	Portable electrical equipment	2.47	Safety harness
2.22	Earth leakage relays: use and check	2.48	Hand protection
2.23	General electrical installations and hazardous atmospheres	2.49	Control over personal protective equipment
		2.50	Notices and signs: electrical, mechanical, protective equipment, traffic signs,

FIGURE 17.4 Shows an example of part of an internal audit report showing the elements, the element number, and the percentage scores.

Weighting Systems

Although traditional audit score weighting systems allocate different points for each SMS element minimum standard detail, the weighting may not be relevant in certain industries and, under certain circumstances, could weaken the intent of the audit process. To weigh each element – minimum standard and minimum standard detail equally, would be an ideal situation. Some SMSs have incorporated a (0–5) rating for each minimum standard detail of the SMS. This rating implies that each minimum standard detail is as important as the next. The scale indicates 0% compliance to 100% compliance (Figure 17.5).

STEP 7 – CORRECT DEVIATIONS FROM STANDARDS

Corrective action is the safety management work that must be done to correct those activities that were not completely controlled. If any critical SMS element is evaluated at <100%, some action needs to be taken to ensure total conformance with standards. Management must do what it says it is going to do. The safety standards indicate what must be done. Deviations indicate that the safety objective has not yet been achieved.

ACCIDENT INVESTIGATION

An accident is caused by a failure in the management system, and after an investigation, certain action plans or controls are recommended to prevent a recurrence of a similar accident. This is corrective action and should be directed at the root causes of the problem, not merely the symptoms. Correcting high-risk behavior and high-risk conditions may provide temporary relief, but the real cause must be identified, and the problem solved.

SMS AUDIT 53 ELEMENTS SECTION 2

OBSERVATION GUIDE	SCORE (1=Low 5=High)					HAZARD	CONDITION COMMENTS	ACTION	
2.00 – Mechanical, Electrical & Personnel Safeguarding	1	2	3	4	5	CLASS A-B-C		BY Who	DATE
2.1 – Machine Guarding (10)									
All hazardous moving machine parts within normal reach completely guarded/fenced									
Inside of guards									
2.2 – LOCKOUT -SYSTEM & USAGE (15)									
Written procedure available and applied									
All equipment can be isolated & locked (including where fuses are drawn or valves in chemical lines, etc.)									
Main switch accessible									

FIGURE 17.5 An example of a (0–5) scoring system for two elements of a SMS.

Fatal Accident Investigation Findings

After the investigation of a fatal accident, the following appeared in the report:

> Control measures in the form of standards, procedures, and monitoring of compliance to procedures are aimed at eliminating the root causes of an accident in an effort to eliminate the unsafe acts and unsafe conditions. Instituting positive control measures will ensure that the same type of undesired event is eliminated. Positive action is required, and ongoing monitoring of these actions is essential.

The results of the same accident investigation indicated that there were some 16 SMS elements (programs), minimum standards, and minimum standard details, which were not controlled. The report concluded that if these control elements had been correctly implemented, the accident would most likely not have occurred.

Common Mistakes

The two most commonly used solutions after an accident are (1) to train workers and (2) to write a procedure. These actions may sound very effective, but they do not necessarily provide a complete cure for the problem. The reasons (root causes) for the deviations from standards should be first identified, then corrective action initiated to fix the problem.

Action plans should be drawn up and they should be specific and contain the following directives:

- What must be done?
- Who must do it?
- By when or how often must it be done?
- When must a reevaluation be conducted to ensure conformance?

SMART

All action plan goals should be SMART, meaning that they should be:

- Specific – The action plan must specify exactly what must be done in detail. It should not be vague or generalize.
- Measurable and manageable – The objective must be measurable and manageable. Unrealistic goals will never be achieved. If it gets measured, it can be managed.
- Achievable and advantageous – The goal must be achievable considering costs and resources. It should be aligned with the organization's objectives and be advantageous to the organization.
- Realistic and result-oriented – Goals must be realistic, and results orientated. A goal such as "injury free" sounds nice is ideal, but is simply not realistic in a workplace setting. A goal such as the holding of safety committee meetings monthly is achievable and realistic.
- Time-related – All objectives should be time-related and have deadline dates for completion. If left open ended, they will not be achieved. The action plan should specify timelines for the tasks at hand and allocate milestones for the achievement of the goal. All goals should be tangible (Figure 17.6).

OBJECTIVE	WHO	TASKS	MEASUREMENT CRITERIA	TIME	RESOURCES NEEDED	FINISHED
1. To complete a safety induction training course.	J. Kane	Write lesson plan and develop visual aids	All new employees and contractors made aware of the site safety rules	Dec. 30	Training room, equipment, visual aids, white board	
2. Implement a near-miss incident reporting system.	P. Roberts	Design a system, develop report form, implement and train employees	Near-miss system up and running, forms developed and reports received. Actions taken	July 1 August 5 Sept 2	Printing and design assistance. Training department to help training.	
3. Inspect all ladders weekly.	Maintenance Supervisor	Checklist, Inspections, corrections	All ladders in a safe condition	By Feb 7	Checklist, training schedule inspections	
4. Hold monthly safety meetings.	Managers	Arrange venue, notify all, compile agenda	Monthly meeting are being held and minutes taken	As from Sept 1st	Meeting room, secretarial assistance	
5. Select, appoint and train Safety Representative.	Section manager and safety department	Select, train and appoint	Sufficient Representatives are appointed and trained. Inspections done monthly	By Feb 23	Appointment criteria, training facilities, Inspection checklist	

FIGURE 17.6   An example of an action plan.

STEP 8 – COMMENDATION FOR COMPLIANCE

One of the main failings in numerous safety processes and programs is the lack of commendation and recognition. Commendation should be given on achievement of objectives. If a department meets and maintains the housekeeping standard, for example, the entire group should be commended. Commendation for pre-contact safety activities is far more effective than commendation for injury-free periods.

PUNISHMENT

The accident sequence indicates a breakdown in the management control system. Weak management controls are normally as a result of failing to assess and manage the risks. Control is a management function and is the most important function to reduce the probability of undesired events, such as accidents, occurring.

All too often supervision resorts to traditional management styles and institutes disciplinary actions in the form of verbal or written warnings or more drastic punitive measures. As is clearly stated in the management principle of *definition,* solutions to problems can only be prescribed once the real cause of the problem has been identified.

COMMENDATION

If, on the other hand, as a result of the measurement and evaluation, a high degree of conformance to performance standards is found, commendation should be given. Safety, as a profession, has often been guilty of emphasizing the lack of control and not complimenting where good control exists.

RECOGNITION

People at workplaces and in other walks of life thrive on recognition and acknowledgment. Recognizing and acknowledging people for safety work done to prevent loss should be done as often as possible. Traditionally, safety recognition is only given to individuals for being "injury free" or for having worked a certain number of days without lost-time injury. That could just be the result of good luck.

Maintaining good housekeeping, carrying out monthly inspections of ladders, portable electrical equipment, hand tools, personal protective equipment, etc., is an ongoing control system and this effort should be recognized.

HEALTH AND SAFETY REPRESENTATIVES

Health and safety representatives have an extremely difficult task in identifying and reporting deviations from standard and having to compile a monthly report. Acknowledging their input recognizes this ongoing effort and meeting of the standard.

MANAGEMENT

It is good management practice to commend employees for their safety effort. Bearing in mind that control is pre-contact accident control, this is more important than the recognition of no adverse consequence in the form of severe injury.

As discussed, positive behavior reinforcement is the key to the success of any SMS. Of all the functions carried out by leadership it is likely to have the most effect on the success of the system. It demands playing the ball and not the man. It requires managers to fix the problem and not the person. It calls for fixing the workplace and not the worker. It forces leaders to deviate from traditional management styles when dealing with issues normally calling for disciplinary measures. After an accident, discipline will not solve the problem and is not a fix-it-all solution. Recognizing good performance will challenge leadership at all levels but will help create more positive leadership across the organization.

CONCLUSION

Health and safety in the workplace is the responsibility of management. Only management has the authority to create a healthy and safe workplace. By implementing management control measures in the form of a risk-based, management-led, and audit-driven SMS, hazards and their associated risks will be identified and reduced, making the workplace safer for all.

18 Health and Safety Management Systems (SMS)

Figure 18.1 shows the domino representing a structured health and safety management system (SMS) which is a stabilizer, and which ensures that the hazards are identified, and risk assessed in order to eliminate accident root causes which lead to the immediate accident causes, the high-risk behaviors, and conditions.

HEALTH AND SAFETY MANAGEMENT SYSTEMS

The best method to stop the accident sequence is to stabilize the three dominos representing the hazard identification and risk assessment process, and the weaknesses in the health and safety management system, which will address the accident root causes. This is done by designing, implementing, and maintaining a structured SMS. The SMS stabilizes the first three dominos and prevents the accident domino sequence occurring.

A SMS is defined as: "on-going activities and efforts directed to control accidental losses by monitoring critical health and safety elements, processes and procedures on an on-going basis." The monitoring includes the promotion, improvement, and auditing of these critical elements regularly. A SMS is the framework of policies, processes, and procedures used to ensure that an organization can fulfill all tasks required to achieve its objectives as defined in the health and safety policy statement.

FIGURE 18.1 The health and safety management system (SMS) forms a stabilizing bridge preventing the accident domino sequence from occurring.

 DOI: 10.1201/9781003385943-21

To guide management in controlling areas of potential loss and to set standards, there are guidelines for SMS that provide excellent frameworks for customization. These systems prescribe certain elements under certain headings and give details of what aspects of an SMS should be instituted as a foundation.

There are at least four crucial areas needed for safety performance. The four key elements are:

1. Top management commitment
2. A humanistic approach towards workers
3. One-on-one contact
4. Use of positive reinforcement

A SYSTEMS APPROACH TO SAFETY

A SMS is a formalized approach to health and safety management through use of a framework that aids the identification, control, and mitigation of health and safety risks. Through routine monitoring, an organization checks compliance against its own documented SMS, as well as legislative and regulatory compliance. It is a series of ongoing management actions, procedures, and processes. A well-designed and operated SMS reduces accidental loss potential and improves the overall management processes of an organization. Introducing a formalized SMS is the only way to change an organization's safety culture and reduce accidental loss.

THE PLAN, DO, CHECK, ACT METHODOLOGY

An SMS can be implemented and maintained by using the *Plan, Do, Check, Act* methodology. A complete safety system would include assignment of personal authority, responsibility and accountability, and a schedule for activities to be completed. Part of a systems approach are the auditing tools to identify and implement corrective actions and thus create a process of sustained continuous improvement.

Another similar process approach, based on the *Plan, Do, Check, Act* methodology, has six steps starting with the health and safety policy, the planning, the implementation, and operation, and after that the checking and necessary corrective action. The last step is the management review of the progress. This cycle leads to a continual safety improvement process.

ONGOING PROCESS

A SMS is a continuous, ongoing process which enables an organization to control its occupational health and safety risks and to improve its health and safety endeavors by means of continuous improvement of health and safety processes. The achievement of targets and goals must be sustained year in and year out. An organization will never be able to state that they have completed the safety management process. As with all processes, existing targets and goals need to be achieved continually and new goals and objectives will arise from time to time.

Risk-Based

The safety system must be a risk-based system. That means it must be aligned to the risks arising out of the workplace. Emphasis on certain SMS elements will be different according to the hazards associated with the work and the processes used. There is unfortunately no safety management system that will be ideal for all mines, industries, and other workplaces, therefore they should be seen as a framework on which to build a risk-specific system for the industry. The main aim of the system is to reduce risks; therefore, the system must be aligned to those risks.

Management-Led

The key factor in health and safety management systems is management leadership. The SMS must be initiated and supported by senior management as well as line management. Only management has the authority and ability to create a safe and healthy workplace. This should be one of their prime concerns.

Health and safety systems that originate in, and which are maintained by the safety department will have little effect on the organization. It is estimated that about 15% of a company's problems can be controlled by employees, but 85% can be controlled by management. Most safety problems are therefore management problems. If managers can manage the intricate and difficult concept of safety, then they will be able to manage other aspects of management easier, as managing health and safety enables them to be more effective.

Audit-Driven

What gets measured usually gets done. Safety is an intangible concept and is traditionally measured after the fact – once a loss has occurred. The SMS must be an audit-driven system which calls for ongoing measurement against the standards and quantification of the results.

Responsibility and Accountability

A SMS converts safety intended actions into proactive activities and assigns responsibility and accountability for those actions. Very similar to what a manager does with his subordinates. Each activity, included in the safety system elements, can be scored on a (0–5) scale to determine whether it has been achieved or not. At the end of the day the entire system can be quantified by the score allocated. The system's effectiveness has been measured. The SMS elements and processes that scored less than full points are highlighted as areas that need improvement. Managers tend to pay more attention to processes that can be measured and quantified, and what gets measured, gets done.

Safety Management Audit Systems

Formal safety management systems were initially seen as audit tools. In reality, an audit is difficult to conduct unless there are standards to audit against, and with no formal safety system in place, audits become baseline measurements rather than measurements of safety management efficiency; hence the terminology, safety audit systems.

HEALTH AND SAFETY MANAGEMENT SYSTEM GUIDELINES

THE NOSA 5-STAR SAFETY MANAGEMENT SYSTEM

The National Occupational Safety Association (NOSA) of South Africa developed the NOSA 5-Star Safety Management System in the early 1970s. This was one of the first audit-based SMSs developed. It was based on 25 years' consulting experience by NOSA field staff and on 150,000 safety surveys that had been conducted across a wide range of industry and mines.

The NOSA 5- Star Safety Management System consists of five sections:

1. Premises and housekeeping
2. Mechanical, electrical, and personal safeguarding
3. Fire protection and prevention
4. Incident (accident), recording and investigation
5. Health and safety organization

Elements

Falling under these five main sections are 73 safety processes, programs, and procedures (control) elements, which constitute the basis of any health and safety management system. The SMS elements contain minimum standards for compliance to each element as well as minimum standard details, which break the element into further achievable and measurable objectives.

THE DNV INTERNATIONAL SAFETY RATING SYSTEM (ISRS)

According to DNV:

> ISRS is a world-leading system used to assess and demonstrate the health of an organization's business processes. Using ISRS gives organizations and their stakeholders' peace of mind that their operations are safe and sustainable.
>
> ISRS is available in 11 languages and has been implemented on thousands of sites worldwide. It includes tried and trusted tools that can be used in flexible ways to develop, train, implement, assess, and benchmark safety and sustainability management.
>
> International Safety Rating System
>
> ISRS represents more than 35 years of accumulated best practice experience in safety and sustainability management. Its continued success is testimony to its vision and strong foundation in research. The first edition was developed in 1978 by Frank Bird, a safety management pioneer, and based on his research into the causation of 1.75 million accidents.
>
> International Sustainability Rating System
>
> ISRS seventh edition was launched in 2006. Its scope went beyond occupational health and safety management to address best practice in environmental, quality and security management and sustainability reporting.
>
> ISRS ninth edition was launched in April 2019. This edition is significantly updated in all areas and describes best practice in 10 loss categories: occupational health, occupational safety, process safety, security, environment, quality, asset integrity, energy, knowledge, and social responsibility.

DNV Website (2022)

THE OCCUPATIONAL SAFETY AND HEALTH ADMINISTRATION, VOLUNTEER PROTECTION PROGRAM

There are a number of national and international guidelines available for the structuring of a SMS. The Occupational Safety and Health Administration (OSHA), Volunteer Protection Program (VPP) is one such SMS guideline:
According to OSHA:

> The Voluntary Protection Program (VPP) promotes effective worksite-based safety and health. In the VPP, management, labor, and OSHA establish cooperative relationships at workplaces that have implemented a comprehensive safety and health management system. Approval into VPP is OSHA's official recognition of the outstanding efforts of employers and employees who have achieved exemplary occupational safety and health.
>
> The Voluntary Protection Program (VPP) recognizes employers and workers in the private industry and federal agencies who have implemented effective safety and health management systems and maintain injury and illness rates below national Bureau of Labor Statistics averages for their respective industries. In VPP, management, labor, and OSHA work cooperatively and proactively to prevent fatalities, injuries, and illnesses through a system focused on: hazard prevention and control; worksite analysis; training; and management commitment and worker involvement.
>
> To participate, employers must submit an application to OSHA and undergo a rigorous onsite evaluation by a team of safety and health professionals. Union support is required for applicants represented by a bargaining unit.
>
> The main elements of the VPP program are:
> *   Management Leadership and Employee Involvement
> *   Worksite Analysis
> *   Hazard Prevention and Control
> *   Safety and Health Training (OSHA Website, 2022)

AMERICAN NATIONAL STANDARDS INSTITUTE

A good guideline for what elements a safety management system should contain is the American National Standards Institute (ANSI), ANSI/AIHA Z10–2005 American National Standard – *Occupational Health and Safety Management Systems*. A similar standard exists for the construction and demolition industry: ANSI/ASSE A10.38–2000 (R2007) *Basic Elements of an Employer's Program to Provide a Safe and Healthful Work Environment*.

OCCUPATIONAL HEALTH AND SAFETY MANAGEMENT SYSTEMS SPECIFICATION (BS OHSAS 18001)

British Standards (BS) OHSAS 18001 is the Occupational Health and Safety Management Systems Specification guideline. Some of the key principles included in the self-assessment questionnaire are:

*   Commitment and policy
*   Planning
*   Implementation and operation

- Support
- Checking
- Audit
- Management review

The BS OHSAS 18001 standard is an international standard which sets out the requirements for occupational health and safety management good practice for organizations. It provides guidance to help organizations design their own health and safety framework.

OSHA 18001 has since been withdrawn and replaced with ISO 45001:2018.

ISO 45001

The International Organization for Standardization (ISO) 45001:2018 – Occupational Health and Safety Management System – Requirements.

ISO has developed an Occupational Health and Safety (OHS) Management System standard guideline (ISO 45001:2018), which is intended to enable organizations to manage their OHS risks and improve their OHS performance. The implementation of an OHS Management System will be a strategic decision for an organization that can be used to support its sustainability initiatives, ensuring people are safer and healthier and increase profitability at the same time.

This standard, rated as the first global health and safety management system, inspired by the well-known OHSAS 18001, is designed to help companies and organizations around the world ensure the health and safety of the people who work for them.

According to the ISO website:

ISO 45001:2018 specifies requirements for an occupational health and safety (OH&S) management system, and gives guidance for its use, to enable organizations to provide safe and healthy workplaces by preventing work-related injury and ill health, as well as by proactively improving its OH&S performance.

ISO 45001:2018 is applicable to any organization that wishes to establish, implement, and maintain an OH&S management system to improve occupational health and safety, eliminate hazards and minimize OH&S risks (including system deficiencies), take advantage of OH&S opportunities, and address OH&S management system non-conformities associated with its activities.

ISO 45001:2018 helps an organization to achieve the intended outcomes of its OH&S management system. Consistent with the organization's OH&S policy, the intended outcomes of an OH&S management system include:

a) continual improvement of OH&S performance.
b) fulfillment of legal requirements and other requirements.
c) achievement of OH&S objectives.

ISO 45001:2018 is applicable to any organization regardless of its size, type, and activities. It is applicable to the OH&S risks under the organization's control, taking into account factors such as the context in which the organization operates and the needs and expectations of its workers and other interested parties.

ISO 45001:2018 does not state specific criteria for OH&S performance, nor is it prescriptive about the design of an OH&S management system.

ISO 45001:2018 does not address issues such as product safety, property damage or environmental impacts, beyond the risks to workers and other relevant interested parties.

ISO 45001:2018 can be used in whole or in part to systematically improve occupational health and safety management. However, claims of conformity to this document are not acceptable unless all its requirements are incorporated into an organization's OH&S management system and fulfilled without exclusion

ISO Website (2022)

ILO-OSH 2001

The document ILO-OSH 2001, entitled, *Guidelines on Occupational Safety and Health Management Systems* was drafted by the International Labor Organization (ILO) in 2001, and the second edition was released in 2009.

The ILO states:

At the onset of the twenty-first century, a heavy human and economic toll is still exacted by unsafe and unhealthy working conditions. The positive impact of introducing OSH management systems at the organization level, both on the reduction of hazards and risks and on productivity, is now recognized by governments, employers, and workers.

The Guidelines call for coherent policies to protect workers from occupational hazards and risks while improving productivity. They present practical approaches and tools for assisting organizations, competent national institutions, employers, workers, and health management systems, with the aim of reducing work-related injuries, ill health, diseases, incidents, and deaths.

At the organizational level, the Guidelines encourage the integration of OSH management system elements as an important component of overall policy and management arrangements. Organizations, employers, owners, managerial staff, workers, and their representatives are motivated in applying appropriate OSH management principles and methods to improve OSH performance.

Employers and competent national institutions are accountable for and have a duty to organize measures designed to ensure occupational safety and health. The implementation of these ILO Guidelines is one useful approach to fulfilling this responsibility. They are not legally binding and are not intended to replace national laws, regulations, or accepted standards. Their application does not require certification.

ILO Website (2022)

EXAMPLE SMS

Section 1 of an Example SMS

Section 1 of the example SMS shows the 30 elements of this section. A comprehensive SMS has five or more similar sections containing other elements which normally total around 70–80 elements. These are broken down into further minimum standards and minimum standard detail in the audit protocol (Figure 18.2).

Number	Element Title	Number	Element Title
1.1	Managers Responsible for Safety and Health	1.16	Safety Newsletters
1.2	Safety Policy: Management Involvement	1.17	Safety and Health Representatives
1.3	Safety Performance Indicators (SPI)	1.18	Safety Management System Audits
1.4	Safety Committees	1.19	External Third Party Audits
1.5	Management of Change	1.20	Safety Publicity Boards
1.6	Safety and Health Training	1.21	Publicity, Bulletins, Newsletters, etc.
1.7	Work Permits	1.22	Safety Competitions
1.8	Organization Risk Management	1.23	Toolbox Talks, Safety Briefings, etc.
1.9	Written Safe Work Procedures	1.24	Safety Specifications:
1.10	Planned Job Observation	1.25	Safety Rule Book
1.11	Safety Inspections	1.26	Safety Reference Library

FIGURE 18.2 An example of a SMS showing the 30 elements of Section 1 of the example SMS. (From McKinnon, Ron C. 2016. *Risk-Based, Management-Led, Audit-Driven, Safety Management Systems* Figure 7.1. Boca Raton, FL: Taylor & Francis. With permission.)

ELEMENT / PROGRAM / PROCESS	POINTS	QUESTIONS THAT COULD BE ASKED	VERIFICATION	WHAT TO LOOK FOR
HEALTH AND SAFETY POLICY				
Policy signed by Executive and prominently displayed in the workplace?	5	What is the company policy?	See the policy	.Check the following requirements
Are employees aware of and familiar with the policy?	5	What is the company policy?	How are employees taught the safety and health policy?	Is it signed by the management? Ask at least 2 employees to name some concept contained in the health and safety policy in their own words.
Is top management committed e.g. attending SMS evaluations and audits, making presentations, etc.?	5	What other involvement does management have in the SMS?	Is the management really involved in the SMS?	Managers partaking in the audit process
Injury / disease prevention	5	Is there a commitment in the policy?	Include a commitment to injury and ill-health and disease prevention	Check policy document
Improvement	5	Is continuous improvement included?	Refer to continuing improvement of safety and health initiatives	Check policy document
Health and safety objectives	5	Are objectives set by the policy?	Provide a framework for safety objective setting	Proactive and reactive measurements
Legal compliance	5	Does the policy state this?	Contain a commitment to comply with safety and health legislation	
Communicated	5	How is the policy disseminated	Be communicated to all affected parties	Induction training/handbook
Displayed	5	Where is the policy displayed?	Be documented, displayed and maintained	Visual confirmation during inspection
Policy updated	5	When was the policy last updated?	Date of update	Check policy issuance
TOTAL	50			

FIGURE 18.3 An example of an audit protocol for the Element – Health and safety policy. (From McKinnon, Ron C. 2020. *The Design, Implementation, and Audit of Occupational Health and Safety Management Systems.* Figure 8.1. Boca Raton, FL: Taylor & Francis. With permission.)

AUDIT PROTOCOL

The audit protocol in Figure 18.3 shows the element of *Health and Safety Policy* as well as the minimum standards required for the element. Each minimum standard is scored on a (0–5) range with the total points for this element being 50. The protocol also includes questions to be asked, documented evidence required, and what the auditor should note during the inspection.

SUMMARY

A structured SMS should be implemented into each workplace to address the risks arising from the processes within that workplace. The SMS is a formalized approach to health and safety management through use of a framework that aids the identification and control of health and safety risks and is the only way to reduce accidental loss from occurring. A SMS stabilizes the weakness in the safety management process and reduces the possibility of accidental loss occurring. It ensures that hazards are recognized on an ongoing basis, that the risks are assessed and that accident root causes are eliminated before they manifest. This will reduce the probability of an accidental loss occurring.

References

Bird, F. E. Jr. and Germain, G. L. (1992). *Practical Loss Control Leadership* (2nd ed.). Loganville, GA: International Loss Control Institute.

Bird, F. E. Jr. and Germain, G. L. (1996). *Practical Loss Control Leadership* (3rd ed.). Loganville, GA: Det Norske Veritas.

Boylston, R. P. (1990). *Managing Safety and Health Programs.* New York: Van Nostrand. © Reprinted by permission of Whiley-Liss, Inc., a subsidiary of John Wiley & Sons, Inc.

Carroll, L. (1978). *Alice's Adventures in Wonderland.* London: Octopus Books.

Columbia Accident Investigation Board (CAIB) Report (2003, August). NASA – Report of Columbia Accident Investigation Board, Volume I.

Compliance Magazine (CM) (1998). Speaking with Jerry Scannell. *Compliance Magazine,* 2. © Reprinted with permission of *Compliance Magazine,* IHS Publishing Group. All rights reserved.

De Ionno, P. and Dlamini, J. (1995, May 14). The level 72 horror. *The Sunday Times,* p. NNEWS9. Johannesburg.

DNV Website (2022). International Sustainability Rating System (ISRS) – DNV.

Ferry, T. S. and Weaver, D. A. (1976). *Directions in Safety.* Springfield: Charles C. Thomas.

Friend, M. A. (1997). Examine your safety philosophy. *Professional Safety,* February, pp. 34–36.

Geller, E. S. (1996). *Working Safe.* Radnor: Chilton Book Company.

Grose, V. L. (1987). *Managing Risk.* Hoboken, NJ: Prentice-Hall.

Health and Safety Executive (HSE) (UK) (1976). *Success and Failure in Accident Prevention.* The Health and Safety Executive HSE (UK).

Health and Safety Executive (HSE) (UK) (1993). The HSE accident ratio.

Health and Safety Executive (HSE) (UK) (1999). Managing human failures: Human factors/ ergonomics – Managing human failures (hse.gov.uk).

Heinrich, H. W. (1959). *Industrial Accident Prevention* (4th ed.). New York: McGraw-Hill Book Company.

Heinrich, H. W., Petersen, D., and Roos, N. (1969). *Industrial Accident Prevention* (5th ed.). New York: McGraw-Hill Book Company.

Howe, J. (1998). A union view of behavioral safety. *Industrial Safety and Hygiene News,* p. 20.

Hoyle, B. (2005). *Fixing the Workplace, Not the Worker: A Worker's Guide to Accident Prevention.* Lakewood, CO: Oil, Chemical and Atomic Workers International Union, pp. 2–4, 7, 10, 17.

Hudson, L. A. (1995) *Insights into Management.* An internal document of Western Deep Levels mine entitled, *A Guide to Effective Accident/Incident Investigation,* p. 2.

Industrial Safety and Hygiene News (ISHN) (2015). Building/sustaining safety cultures: EHS pros' top priority for '15, says ISHN survey | 2015-01-15.

International Labour Organization (ILO) (2001). *Guidelines on Occupational Safety and Health Management Systems, ILO-OSH 2001* (2nd ed.). Geneva: ILO.

International Organization for Standardization (ISO) (2022). ISO 45001:2018, Occupational health and safety management systems: Requirements with guidance for use.

Johnson, D. (1998a, September). Cooking the books. *Industrial Safety and Hygiene News,* p. 21.

Johnson, D. (1998b). What does management think about safety and health? *Industrial Safety and Hygiene News:* 15th White Paper Survey, p. 22.

Krause, T. R. (1997). *The Behavior-Based Safety Process* (2nd ed.). New York: Van Nostrand Reinhold. © Reprinted by permission of Whiley-Liss, Inc., a subsidiary of John Wiley & Sons, Inc.

Manning, M. V. (1998). *So, You're the Safety Director!* (2nd ed.). Rockville: Government Institutes, Inc.

Manuele, F. A. (1997). *Professional Safety*, p. 31.

McKinnon, R. C. (1992). *Safety Management*, September 1992, p. 12.

McKinnon, R. C. (1995). *Five Star Safety: An Introduction.* Unpublished work.

McKinnon, R. C. (2012). *Safety Management, Near Miss Identification, Recognition, and Investigation,* (Model 2.2, 2.4). Boca Raton, FL: Taylor & Francis.

Mine Safety and Health Administration (MSHA) (2022). *MSHA - Performance Coal - Upper Big Branch Mine-South - Fatal Accident Report - Executive Summary.*

National Occupational Safety Association (NOSA) (1988). *Advanced Questions and Answers* (Vol. HB5.11E). Pretoria: National Occupational Safety Association.

National Occupational Safety Association (NOSA) (1990). *Effective Accident/Incident Investigation* (Vol. HB4.12.50E). Pretoria: National Occupational Safety Association.

National Occupational Safety Association (NOSA) (1994). *Health and Safety Training* (Vol. HB5.31.55E). Pretoria: National Occupational Safety Association.

National Occupational Safety Association (NOSA) (1995). *The NOSA 5 Star System.* (Vol. HB0.0050E). Pretoria: National Occupational Safety Association.

National Safety Council (NSC) (1993). *Accident Facts 1993* (Vol. 1993). Itasca: National Safety Council.

National Safety Council (NSC) (2020). *Injury Facts 2020,* Injury Facts - National Safety Council (nsc.org) Permission to reprint/use granted by the National Safety Council © 2021.

National Safety Council (NSC) (2022) *Injury Facts 2020,* Work Safety Introduction - Injury Facts (nsc.org) Permission to reprint/use granted by the National Safety Council © 2021.

Occupational Safety and Health Administration (OSHA) (2009). Voluntary Protection Programs | Occupational Safety and Health Administration (osha.gov)

Occupational Safety and Health Administration (OSHA) (2022). Top 10 Most Frequently Cited Standards | Occupational Safety and Health Administration (osha.gov).

Petersen, D. (1978). *Techniques of Safety Management* (2nd ed.). New York: McGraw-Hill Book Company.

Petersen, D. (1988). *Safety Management.* New York: McGraw-Hill Book Company.

Petersen, D. (1996). *Analyzing Safety System Effectiveness* (3rd ed.). New York: Van Nostrand Reinhold. © Reprinted by permission of Whiley-Liss, Inc., a subsidiary of John Wiley & Sons, Inc.

Petersen, D. (1998). What measurement should we use, and why? *Professional Safety,* pp. 37–39.

Prickett, J. (1998). Incentive programs reflect management's attitude. *Industrial Safety and Hygiene News,* pp. 12–13.

Simonds, R. H. and Grimaldi, J. V. (1963). *Safety Management* (8th ed.). Homewood: Richard D. Irwin, Inc.

Smit, E. and Morgan, N. I. (1996). *Contemporary Issues in Strategic Management* (1st ed.). Pretoria: Kagiso Tertiary.

Smith, S. L. (1994). Near-misses: safety in the shadows. *Occupational Hazards,* pp. 33–36.

Smith, T. A. (1998). What's wrong with safety incentives? *Professional Safety,* p. 44.

Tarrants, W. E. (1980). *The Measurement of Safety Performance.* New York: Garland STPM Press.

United Steel Workers International Union (2005). http://assets.usw.org/resources/hse/resources/Walking-the-Talk-Duponts-Untold-Safety-Failures.pdf.

Index

Note: *Italic* page numbers refer to figures.